SpringerBriefs in Astronomy

More information about this series at http://www.springer.com/series/10090

Timothy F. Slater · Coty B. Tatge

Research on Teaching
Astronomy
in the Planetarium

 Springer

Timothy F. Slater
University of Wyoming
Laramie, WY
USA

Coty B. Tatge
University of Wyoming
Laramie, WY
USA

ISSN 2191-9100
SpringerBriefs in Astronomy
ISBN 978-3-319-57200-0
DOI 10.1007/978-3-319-57202-4

ISSN 2191-9119 (electronic)

ISBN 978-3-319-57202-4 (eBook)

Library of Congress Control Number: 2017940535

Printed on acid-free paper

This Springer imprint is published by Springer Nature
The registered company is Springer International Publishing AG
The registered company address is: Gewerbestrasse 11, 6330 Cham, Switzerland

Contents

Chapter 1
Astronomy Education Research in the Planetarium

From the first moment intervening clouds obscured the beauty of the overhead stars, our ancestral sky watchers created pictures and maps to represent the heavens. For some, precise star positions were needed for navigation and wayfinding. For others, knowing which bright stars were visible during which season guided scheduling of religious ceremonies or signaled when planting might most fruitfully occur. Even in modern times, people are still interested in the symphony of celestial motions overhead. Mobile smart phones can readily carry simulations of the sky, empowering even the most casual observer to quickly identify the position of the setting Sun, the name of the bright planet shining overhead, or the constellations grazing the southern horizon. Such a longstanding and ongoing fascination with knowing the sky—either academically or spiritually—motivates astronomy educators to build monument-like museums as altars to the stars. Thus was born the planetarium.

Summarized eloquently by Petersen (2003), the history of planetarium education is rich. In the 1930s, enthusiasm for installing planetariums, or planetaria, was feverish. In the United States, awe-inspiring planetariums were installed first in Chicago (Adler Museum), to be quickly followed by planetariums in Los Angeles (Griffith Observatory) and New York (Hayden at the American Museum of Natural History). Planetarium facility construction increased rapidly in the United States due to federal funding to schools and museums through the 1958 US National Defense Education Act, and the US went from one planetarium in 1930 to many hundreds today. An annually updated and detailed catalog of planetarium locations can be found through the *International Planetarium Society's* website (Planetariums of the World 2016). Looking toward the future of planetariums, we find a field of education that is driven by a combination of advances in technology (Wyatt 2005), a desire to communicate everything from basic astronomy to the latest discoveries in astrophysics and an ongoing passion of planetarium educators to save the sky for future generations (Reed 1989).

Since that first planetarium installation nearly 100 years ago, researchers in education have grappled with understanding how people learn in the planetarium. Scholarly efforts to quantify just how much learning occurs in the planetarium

© The Author(s) 2017
T.F. Slater and C.B. Tatge, *Research on Teaching Astronomy in the Planetarium*,
SpringerBriefs in Astronomy, DOI 10.1007/978-3-319-57202-4_1

created a new academic field within astronomy and science education research—the discipline of *planetarium education research*. The numerous doctoral dissertations, refereed journal publications and professional conference proceeding presentations, have resulted in hundreds of scholarly planetarium education research reports that describe methods and results.

There are only a few comprehensive, scholarly literature reviews published in academic journals that cover the entire landscape of planetarium education research, and none within the last decade. The lack of scholarly literature reviews is not evidence of an immature scholarly field; instead, only a few literature reviews have been published because planetarium education research is published across a far reaching range of scholarly venues making gathering and synthesizing existing research an arduous, if not Herculean-level, task. In fact, the most comprehensive literature reviews of planetarium education research existing are not actually found in refereed-journal publications at all, but instead located within the extensive literature review sections of doctoral dissertations. Until now, dissertations and graduate theses serve as the majority of academic sources that review the field's portfolio, generally in the format of an annotated bibliography and often leveraging other dissertations concerning planetarium education, rather than relying on referred journal publications or professional conference meeting proceedings as foundational sources.

The existing peer-reviewed literature summarizing the results and methods of planetarium education research are listed in Table 1.1. There are only five standalone reviews to date that are exclusively related to the planetarium education research literature. Each one of the authors of these reviews took radically different structural and philosophical approaches on how they designed, analyzed, and synthesized the literature reviewed. Specifically, they vary across four different domains: (*i*) content scope of what is being reviewed; (*ii*) geographic scope; (*iii*) types of documents reviewed; and (*iv*) the analysis methods used. The category *content scope* describes the range and domain of astronomical concepts included in

Table 1.1 Summary of peer-reviewed literature reviews in planetarium education

Author	Year	Document types	Analysis method	Number of studies included
Smith	1974	Peer-reviewed papers, dissertations and grey literature	Annotated bibliography	>100
Sunal	1976	Peer-reviewed papers and dissertations of empirical studies involving student outcomes	Descriptive statistics	9
Riordan	1991	Peer-reviewed papers and dissertations	Annotated bibliography	$\cong$25
Hunt	1993	Peer-reviewed papers and dissertations	Annotated bibliography	$\cong$20
Brazell	2009	Peer-reviewed papers and dissertations	Quantitative meta-analysis	19

the paper. *Geographic scope* subdivides neatly into two thematic groups: those reviews that only included works in the US and those reviews that included works in the larger international setting. *Document types* is a category describing the way in which the authors chose to restrict the sources for works they included in their reviews. Finally, the category of *analysis methods* describes the approach the authors used to consider the works they reviewed. Note that few of the sources listed within these reviews do not explicitly state the scope, type, analysis method, or number of studies included. Overall, for a scholarly literature review to be judged as having used a synthesis effort as an analysis method, multiple literature sources must be evaluated, compared, and contrasted to infer actionable guidance for the future planetarium education research endeavors. Descriptive reviews, on the other hand, use analysis methods that are often quantitative but exclude any type of inferences and/or recommendations regarding the future of the field. An annotated analysis method pertains to how literature is presented, where each piece of literature is summarized in paragraph form. And lastly, each literature review can be selective in the scope and selection of literature included in the analysis.

The first published peer-reviewed literature review of planetarium education was compiled and evaluated by Smith (1974). Considered to be highly comprehensive for its generation, more than 100 literature sources were included in Smith's scholarly review. A multitude of sources were covered, including theses and dissertations, peer-reviewed journal articles, and grey literature works related to the planetarium. Grey literature works are those including professional meeting presentations, conference proceedings, and society and association newsletter articles and other resources not generally considered scholarly, peer-reviewed academic reports (Slater 2015).

All the works reviewed by Smith (1974) described planetarium education research efforts conducted within the United States (US) with the exception of works published in the *Planetarian*, a journal run by the *International Planetarium Society*, which includes articles from an international scope. Each of these sources was presented as an annotated bibliography. Smith divided the field into three categories of works: descriptive, comparative, and curriculum studies. Descriptive studies were considered to be those that "attempted to describe the status of planetarium operations at various stages of development." Comparative studies were empirical comparisons of the effectiveness of classroom and planetarium, typically with learning astronomical concepts. Curriculum studies were focused on which grade levels with the best time to teach particular astronomical concepts within the planetarium. Smith (1974) described each study in an annotated bibliography format, and then reported his findings as:

(1) There are conflicting results for the value of the planetarium.
(2) There is a need to determine which concepts are most advantageous in the planetarium for respective grade levels.
(3) Planetarium is effective for low-socioeconomic status students.
(4) There is a conflict in results of whether the planetarium is most effective in cognitive or affective domain.

(5) Need to determine frequency of planetarium visits for effectiveness.
(6) Evaluation of planetarium apparatuses that are most valuable to enhancing particular astronomical concepts is needed.
(7) More research is needed on live versus taped lectures.
(8) More coordination with a variety of facilities to contribute to the planetarium curriculum.
(9) More standardized assessment instruments are needed for the planetarium.

Smith was able to exceptionally synthesize the field and gave planetarium education researchers direction for future studies.

A few years later, Sunal (1976) published the review article, "Analysis of Research on the Educational Uses of a Planetarium" in which he created a theoretical model to situate planetarium research. The model he advocated consisted of classifying planetarium studies into three domains: cognitive, affective and process skills scheme. Sunal classified nine out of the total 16 studies he found from a literature search related to planetariums from 1946 to 1976. The criterion for the studies to be included was that it needed to measure student outcomes from their experience in the planetarium. Sunal used descriptive statistics to conduct his analysis of peer-reviewed works and dissertations from the US, all of which pertained to empirical studies of student outcomes. He then further divided the works to analyze by education level, such as elementary school, secondary school, and college/university higher education. Sunal (1973) proposed the following conclusions:

1. "The planetarium experience can change student performance in most, if not all, of the goal areas of planetarium educators.
2. Research does not support the use of the traditional one-visit planetarium unit. With regard to performance on objectives, more can be done in a classroom lesson.
3. As part of an instructional unit, a planetarium lesson is as effective as a classroom lesson covering the same objectives in changing student performance in the cognitive and process-skill areas, the benefit of a planetarium lesson to a student appears in producing change in the affective area, if in any area.
4. Effective strategies for producing greater student performance than traditional planetarium use are combined planetarium-classroom instructional units, use of the planetarium for remedial lessons, visits scheduled later in the classroom unit, and orientation visits.
5. Student differences and strategies which do not appear to be important in effecting gains in performance in a planetarium lesson are sex of the student, IQ, reading and mathematics ability, and pre-planetarium or follow-up exercises."

Sunal's study was intended to develop a framework in which planetarium education researchers could discuss findings and research. When discussing his own findings from the literature search, he presented them in a comparative way and from within the bounds of his own model.

Fifteen years later, Riordan (1991) reviewed planetarium education projects such as classroom projects, planetarium and participatory lessons and focused research. He divided the literature based on learning theory, classroom/planetarium research, participatory oriented research, focused and other research. Riordan used approximately 25 studies in his analysis of peer-reviewed articles and dissertations from the US. Each of these sources was presented as an annotated bibliography. In reference to learning theories, Riordan discussed Bishop's (1976) study which links students' abilities to learn particular astronomical concepts based on age (e.g. students below the age of seven years of age are unable to learn lunar phases and constellations) and Schatz's (1976) study of how a participatory oriented planetarium show can be used to help students entering the concept exploration phases of the Karplus Learning Cycle. While discussing classroom versus planetarium learning environments, Riordan references Wright (1968), Reed (1973), Reed and Campbell (1972), Soroka (1967), Tuttle (1967), Rosemergy (1967) Reed (1970), and Smith (1966). He also mentions how Dean and Lauck (1972) and Warneking (1970) are concerned with the evaluation techniques of these studies. The planetarium research field then shifted towards planetarium program instruction, specifically student-centered participatory oriented programs that included active-learning aspects being advocated by the broader science education community of the day. Studies such as Sneider, Eason, and Friedman (1979); Mallon and Bruce (1982), and Fletcher (1980) are examples of these. He then described studies which focused on particular learning factors (e.g. spatial reasoning skills). Studies included were Sonntag (1988), Sonntag (1989), and Tuttle (1965). The few other studies included were Ridky (1975), Hoff (1970), and Sunal (1976). Riordan encouraged the use of Sunal's model for evaluating the planetarium. Using an annotated bibliography analysis method, Riordan described the planetarium education field using a limited collection of studies that he deemed essential to the planetarium education research community.

Two years later, Hunt (1993) reviewed research of visuals and media from educators and extrapolated them into a planetarium setting. Specifically, he reviewed the visual literacy research, cueing techniques, audiovisual research, and facility goals and objectives. Approximately 20 sources of dissertations and peer-reviewed works within the scope of the US were included in Hunt's analysis. Hunt (1993) proposed the following research questions as being critical to the next steps in the field based on his analysis of the existing literature:

(a) Do planetarium visitors learn more by showing fewer visuals and showing them longer?
(b) Under which format do planetarium visitors learn best, simple black and white line drawings, color photographs or false color photos?
(c) Are cues such as labels (words), color coding of line drawings, color coding of line drawings, arrows, and lines or visuals without cues significant to learning in the planetarium?
(d) Is animation significant for the learning of the planetarium visitor? If animation is significant, does it need cues?

(e) Is animation better than a dissolve sequence showing the same dynamic process? If dissolve sequences are more significant, do they need cues?
(f) What are the values of the music and other planetarium aesthetics?
(g) In an all-sky movie experience, do visitors learn more if visual cues are projected with slide projectors over the movie to distinguish important points in the soundtrack or without cues?
(h) Does the image size make a significant difference in learning in the planetarium?
(i) Do multiple images in a multiscreen format compared to a single screen affect learning in the planetarium?
(j) How do these factors work with visitors of various Piagetian frames of reference?

Hunt was highly selective with the types of articles he reviewed in that he constricted his analysis to works related visuals and media within the planetarium. He described each study in an annotated bibliography format, and then inferred how those studies' results could be transposed or tested in a planetarium setting. Hunts work is notable for giving the planetarium research community explicit direction. It is also notable for using studies that have already been conducted and standing on the shoulders of giants, rather than treating learning with media as an isolated island of research.

Nearly twenty years after Hunt's analysis, Brazell (2009) reviewed "Planetarium Instructional Efficacy: A research synthesis" for his doctoral dissertation at Texas A&M University. Of 46 total studies he located within the scope of the US and that were related closely to planetarium efficacy, only 19 satisfied his criteria to be included in a meta-analysis study. He considered the statistical effect sizes concerning student achievement and student attitudes, which highly constrained which studies he could use for analysis. Using standard practice, quantitative meta-analysis methods, Brazell's dissertation emphasized a selective group of planetarium education research articles related particularly to planetarium efficacy. In order for the studies to be included in calculating the effect sizes, they had to meet the following criterion:

1. The study must involve the use of a planetarium in an instructional capacity.
2. The study must be a comparison between the use of a planetarium and another instructional approach.
3. The study must report outcome measures in terms of student achievement, student attitudes, or both.
4. The study's participants must be students in an instructional setting such as a K-12 school, college, or university that is utilizing the planetarium as a teaching tool.
5. The report must contain sufficient statistical information that will allow the researcher to calculate a suitable effect size (Lipsey and Wilson 2001).
6. The study must employ an experimental or quasi-experimental design utilizing a control group.

7. Both published and unpublished research reports are eligible for inclusion as long as the other selection criteria are met.
8. The study must have been conducted between 1960 and 2008.

Brazell found that "the planetarium has not been a very effective tool for improving student attitudes towards astronomy." However, the planetarium has been statistically effective, if not small in effect size for student achievement.

The highly limited number of reviews of planetarium education research suggests that the planetarium education research community would benefit from a central platform with a collection of all works internationally related to planetarium education. All of the initial literature we have found, relating to planetarium education, was found through an internet search through databases. The databases searched were Google Scholar, University of Wyoming Libraries, WorldCat, ProQuest, and ERIC. Keywords such as "planetarium", "education", "dome", "fulldome", "informal astronomy", and so on were used to search the databases. We collected any articles from peer-reviewed journals, preferably searching for empirical research within the planetarium. Dissertations, theses, and grey literature conference proceedings were also included, if found. Using the works from the initial search, the top 25 peer-reviewed journal websites with the most frequent return of relevant articles were visited and scoured for more relevant works. Again, using the initial collection, each of the works' references was used to deepen the breadth and scope of pertinent articles. Additionally, planetarium education dissertations and theses from a previous collection of astronomy education research were included to the planetarium education research collection. Any dissertations or theses that could be located via Interlibrary Loan as hard copies were ordered, scanned, categorized, and uploaded to the database, and were included in our results.

In order to constrain the wide range of planetarium education research results we've found, we focused our efforts on research reports that occurred in a physical full dome setting, including portable domes. Using this definition, simulated planetariums on 2-D platforms, such as desktop computer software programs that simulate planetariums like *Stellarium*, are not included in our results. It is important for planetarium education researchers to clearly distinguish which contexts they are working, due to the highly varied learning environments in which they can occur. As but one illustrative example, prolific planetarium education researcher Plummer (2006) is well-renowned in the field of planetarium education today even though all of her research does not directly constitute as being planetarium studies. Her dissertation study created an artificial dome to "resemble a very small planetarium" and used a flashlight to imitate apparent motion. (Plummer 2006). Although this type of instructional tool could be considered a variation of the planetarium, it is not the type of studies we included in our review. Each of the planetarium education research works we've reviewed are categorized using the system devised by Slater et al. (2016) and colleagues as shown in Table 1.2.

Due to the surprising lack of empirical research reports in scholarly peer-reviewed journals related to planetarium education research, much of the research is found within grey literature (Slater 2015). Additionally, one of the

Table 1.2 Categories for reviewing planetarium education research reports

Category	Definition
Document source	
Peer-reviewed	Published academic or scientific work judged as valuable, timely, and rigorous by other scholars in the field
Dissertation or Thesis	A paper that is used to fulfill the requirement of a higher educational degree, juried by university faculty
Grey literature	Research presented at conferences, on websites, in newsletters produced by organizations or societies but not through commercial channels, e.g., pre-prints, conference proceedings, statistical reports, preliminary progress reports, etc.
Type of resource	
Literature review	A comprehensive comparison and contrast paper which critically summarizes or synthesizes the current knowledge across an academic landscape
Empirical research	A scientific study conducted using data collected via observation or experiment
Theoretical research	A study describing a pre-existing, redefined, or new theory that makes a priori predictions about proposed experiments
Curriculum or program evaluation	An assessment on the effectiveness of an education program or curriculum design
Curriculum description or report	A paper describing and documenting an educational program or curriculum design
Position paper or editorial	A non-refereed, opinion-oriented publication
Historical	A paper describing past events and its influence today
Bibliography or resource guide	A summary of the literature or annotated guide to existing literature or educational resources
Empirical methodology	
Quantitative	Research designed to quantify an educational state, or change of state, by transforming observations into numerical data that can be statistically analyzed
Qualitative and interpretive	Research designed to explore underlying causal mechanisms occurring and meaning-making participants experience interpreted without extensive use of traditional statistical analysis
Mixed-methods	A study using both qualitative and quantitative methods to triangulate an evidence based conclusion from multiple philosophical perspectives and viewpoints

(continued)

Table 1.2 (continued)

Category	Definition
Learning environment	
Formal	Context is school-like classroom setting, led by trained instructors, with specified and assessed learning targets
Informal	A study conducted outside of a classroom setting (e.g., after-school programs, museums, star parties, field trips, etc.)
Research setting	
Planetarium	Any study incorporating the use of a physical planetarium
Museum	Science learning center or facility that preserves historical artifacts
Amateur groups or activities	A study involving non-professional events (e.g., star parties, variable star data collected via amateur astronomers, etc.)
Extracurricular/camps/after-school/scouting	Non-school related educational activities that are rarely graded
Home school	Students who do not attend a traditional brick-and-mortar school building
Online, virtual	Students who do not attend a traditional brick-and-mortar school building and take synchronous or asynchronous classes via an Internet connection
Citizen science	Individuals who are not paid professional scientists and are contributing to scientific data collection or analysis
Research facility	National laboratory or scientific research institution that is not a college, university or museum
Study participants	
Early childhood students	Younger than 5 years of age
Elementary students	Students ages 5–11 years of age
Middle and secondary school students	Students ages 11–18 years of age
College students	Students enrolled in an undergraduate course
Second language learners	Students enrolled in a course or school not taught in their first language learned
Graduate students	Students pursuing higher education degrees such as a doctorate or masters
Pre-service teachers	College students pursuing a career as a teacher
In-service teachers	Teachers currently educating students aged 3–18 years old
College faculty	Participants teaching at a higher education institution
Multi-aged groups	Family or community groups with varying ages
Adult learners	Retirees or third-age learners (e.g., Elderhostel)

(continued)

Table 1.2 (continued)

Category	Definition
Construct	
Content knowledge	Facts, concepts, theories, etc.
Affective variables	Belief, identity, motivation, attitudes, etc.
Cognitive processes	Attention, memory, working memory, judgment, reasoning, etc.
Quantitative reasoning	Ability to read and understand mathematical problems
Spatial reasoning	The capacity to visualize and manipulate mental images
Research and methods assessment	The research methodologies used in empirical and evaluation studies
General teaching strategies	The didactic, pedagogical, praxis, and educational engagement process
Scientific inquiry	The processes involved in developing scientific knowledge
Nature of science	The understanding of science as a way of knowing
Scientific content focus area	
General/broad knowledge of astronomy content	
Stars	
Sun-earth-moon (includes Seasons and lunar phases)	
Solar system	
Scale and structure	
Cosmology	
Atoms and light	
History of astronomy	
Astrobiology and exoplanets	
Galaxies	
Gravity	
Cultural astronomy	
Other	
Demographic focus	
Multicultural or indigenous	
Gender and sex differences	
Disability	
Gifted	
Low socio-economic status	
Other	

(continued)

Table 1.2 (continued)

Category	Definition
Language of publication	
Paper written in english	
Abstract only in english	
Non-English publication	
Location study conducted	
USA	
Other	

journals central to planetarium education research, the *Planetarian*, publishes both peer-reviewed and grey literature mixed together without a clear distinction between the two. In order for a work to qualify as "peer-reviewed," reports reviewed required an abstract, a listed accepted/published date, and one or more sections considered a methodology, literature review, or program evaluation with empirical results. For a research report to be classified as "grey literature", it needs to fulfill one or all of the following criteria: no abstract, no bibliographic citations, a conference proceeding, poster, or presentation. Since a large portion of publications relating to planetarium education research usually describe the subject in a general manner, they were categorized within the program/curriculum report or description domain as long as they were not considered formal reviews of the literature. Also classified within this category were any works that described a new activity or planetarium.

When considered in total, slightly more than half of the planetarium education research reports reviewed come from traditional, peer-reviewed articles and dissertations (Fig. 1.1). Of the planetarium education research studies included in our review, we decided to focus primarily on empirical studies that could actually inform the community about how learning works and the best practices within the context of the planetarium. There were 74 dissertations and graduate theses related to the planetarium, and 48 of those were empirical studies. The majority of other studies from our search fell under categories such as curriculum or program evaluations, curriculum or program descriptions. For any other works related to planetarium education, as of yet, there are a total of more than 125 journal articles. Of those, only 34 peer-reviewed and four grey literature works qualified as empirical research. In total, there are 86 empirical research articles included in our results, most of which are dissertations and theses (56%) followed by peer-reviewed (40%) and grey literature works (5%) (Fig. 1.1). It is unusual for a field to produce more dissertations or theses than it does actual peer-reviewed empirical articles. For example, for fields such as English, History, and Political Science, only approximately 26% of doctorate recipients never go on to publish in the field (Ehrenberg, et al., 2009, pg. 208). Due to this anomaly in the planetarium education field, it is important that dissertations and theses are considered foundational research from which the field can grow. Dissertations and theses can often times be equally, or more, rigorous as the peer-review process. These types of studies can take years to

Type of Document

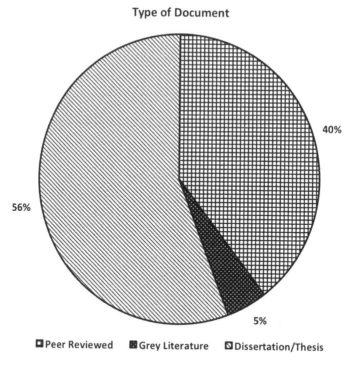

Fig. 1.1 Types and percentages of planetarium education research reports included in this review

complete and are under close inspection and scrutiny of their committee, who are usually experts in the fields themselves.

Each planetarium education research study using empirical research employs one of the three research methods: quantitative, qualitative/interpretive, or mixed-methods (Plummer et al. 2015; Slater et al. 2015). In total, most of the studies were done using quantitative methods (73%, n = 61), followed by qualitative (15%, n = 13), and least common mixed methods (12%, n = 10). As shown in Fig. 1.2, most studies for both dissertations and peer-reviewed works are quantitative (n = 22). However, for peer-reviewed works there are fewer qualitative studies (n = 3) conducted in the planetarium than there are of researchers using mixed methods techniques (n = 5) to collect and analyze their data. It is expected that most research done within the planetarium is conducted using quantitative methods considering most studies analyze the effects on learning gains using a variety of teaching techniques within the dome.

Figure 1.3 shows a bimodal shape of publications of empirical studies in planetarium education research over time, first peaking in the late 1970s (n = 13) and then again in the late 2000s (n = 17). It is important to note that the majority of empirical research studies were conducted as planetarium research projects before 2005 and since then the majority have been published as peer-reviewed works. Again, this data emphasizes the importance of including dissertations and graduate

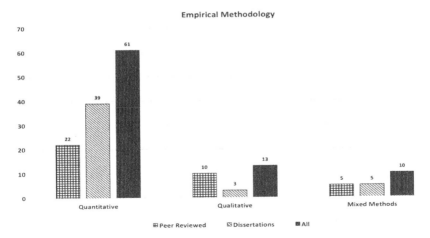

Fig. 1.2 Number of planetarium education research reports by Empirical research methodology

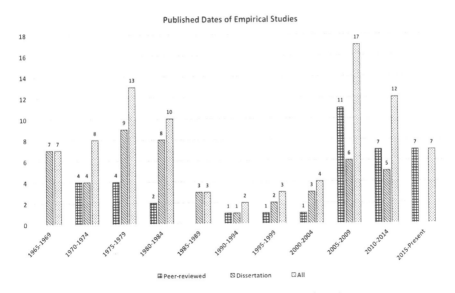

Fig. 1.3 Number of planetarium education research reports in various date ranges

theses related to planetariums as part of reviews of the literature. During 1990–1994, only one dissertation and one peer-reviewed work were published. Around that same time, digital planetariums were just starting to be built, so research in the planetarium was more or less put on hold until the nature of what was possible within a digital dome was better understood. We can divide the planetarium research into two eras: analog and digital (Figs. 1.4 and 1.5, respectively). These two eras also show a change in the dominance of documents, the analog era was

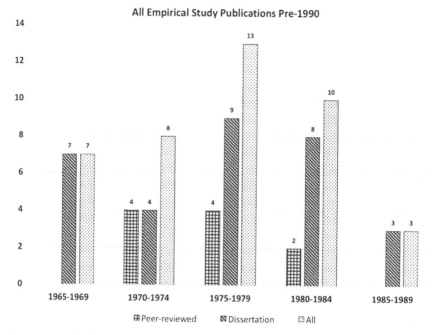

Fig. 1.4 Empirical planetarium education research reports prior to 1990

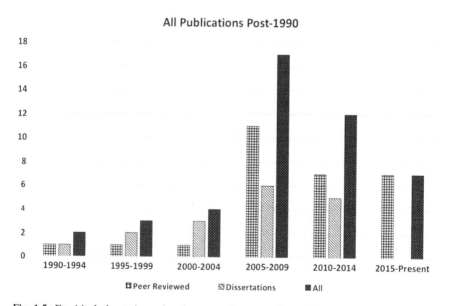

Fig. 1.5 Empirical planetarium education research reports since 1990

mostly dominated by research done by dissertations, and the digital era by peer-reviewed articles.

Over time, changes in the methodologies used in the analog versus digital era were also apparent, as illustrated in Figs. 1.6 and 1.7. Until the late 1980s all empirical research in planetarium education research was conducted using quantitative methods of analysis. This change occurred after the period between 1980 and 2000 which was widely known in the education research community as the time of the "paradigm wars" (Gage 1989). After this period, the broader education research community and its scholarly journal editors agreed that non-experimental and non-quantitative methods were acceptable as legitimate research practices. During the paradigm wars and since, qualitative research tends to be, but is not always defined as, research that has the contextually significant experience of the participants at its focus (Erickson 1985). This is a perspective which had in the past frequently excluded papers designated as qualitative or interpretive in its research methods. The methods used by planetarium education researchers in its early history compared to more modern efforts reflects this history by incorporating qualitative analysis into quantitative studies. It was not until the late 2000s that planetarium education researchers began relying solely on the newer qualitative and interpretive methods for research.

Planetarium education researchers, for both dissertation and peer-reviewed studies, have largely been interested in carefully measuring the acquired content knowledge of students attending planetarium shows (44%, n = 65), thereby establishing the value and effectiveness of planetariums. If teaching and learning inside the planetarium were shown to be equally or more effective than the classroom, this would justify the hundreds of planetariums built and purchased in the middle and late 1900s with the 1958 US National Defense Education Act funds that

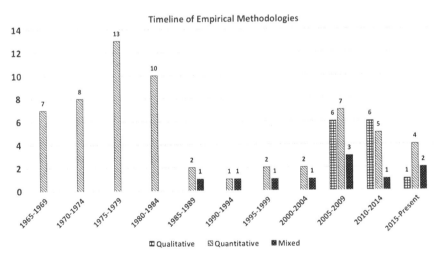

Fig. 1.6 Timeline of evolving empirical research methods in planetarium education research

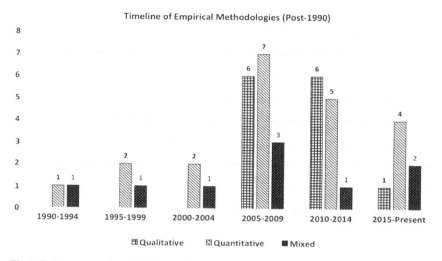

Fig. 1.7 Frequency of empirical research methods in planetarium education research since 1990

were ongoing for more than 30 years. Best practices in contemporary, student-active teaching strategies in the planetarium is also one of the most often measured constructs in planetarium education (21%, n = 31), followed by cognitive processes (7%, n = 11); spatial reasoning (7%, n = 10); perspective (1%, n = 2); and 8 other studies each researching a variety of constructs not represented in Fig. 1.8: Visualizations, Mystic and Seductive Details, Scientific Inquiry, Night-Sky Attachment, Nature of Science, Learning Trajectory, Language, and Humor (1%, n = 1).

Prior to 1990, the field of planetarium education research was dominated by empirical research studies through dissertations and graduate theses (n = 28). The most prolific academic institutions supporting doctoral dissertations related to planetarium education research were the University of Virginia (n = 3), University of Michigan (n = 3), Michigan State University (n = 3), Wayne State University (n = 2), and Pennsylvania State University (n = 2). Fifteen additional universities supported a single dissertation in planetarium education research.

After 1990, the diversity of universities interested in planetarium education research widened rapidly: University of Michigan, Walden University, Virginia Polytechnic Institute and State University, University of Washington, University of Missouri, University of California, University of Auckland, Universite de Montreal, University of Maine, Texas A&M University, Temple University, Northern Illinois University, Indiana University, East Tennessee State University, and Dalarna University. The University of Michigan supported two dissertations and theses, whereas all other institutions only represented one dissertation or thesis each. Considering that the majority of empirical research studies post-1990 were peer-reviewed, the magnitude of dissertations related to planetarium education was simply too small to reflect the impact of research in the field.

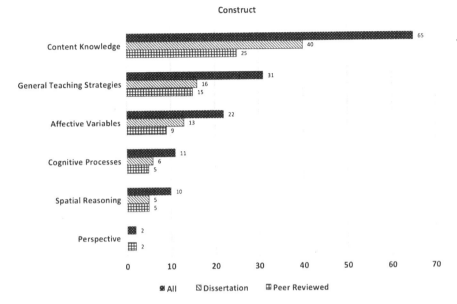

Fig. 1.8 Number of planetarium education research reports by academic construct

Returning to the planetarium education research era prior to 1990, the majority of studies were done as graduate theses or doctoral dissertations and, as common practice during this period, very few doctoral dissertations were published through the peer-review process as articles. Of the total 10 publications, the majority of peer-reviewed publications of empirical studies came from West Chester State College (n = 5), where George Reed served as a faculty member (Fig. 1.12). The other institutions that published on planetarium education research included Arnim D. Hummel Planetarium, the Board of Education of Harford County, Eastern Kentucky University, Fernbank Science Center, Methacton School District, Temple University, University of Maryland, and University of Pennsylvania, in all of which, only one author produced one publication representing that institution (Fig. 1.9).

Of the total 49 peer-reviewed publications in the digital era, Arcadia University (n = 6), Brigham Young University (n = 4), Pennsylvania State University (n = 3), Ohio State University (n = 3), University of Virginia (n = 2), Harvard University (n = 2), and Denver Museum of Nature & Science (n = 2) were the top institutions to be represented multiple times by authors of peer-reviewed works. All other 27 institutions only had one author representing a publication from that institution. Arcadia University, which was most represented within the 49 publications, is where Julia D. Plummer served as a faculty member until 2012 before she took an opening as an associate professor at Pennsylvania State University, another institution largely represented in the planetarium education research field.

Over time, the nature and demographics of planetarium education research study-participants has evolved. Again, there is a dramatic difference prior to 1990

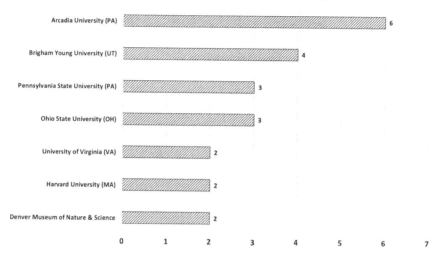

Fig. 1.9 Peer-Reviewed publications per institution (Post-1990)

(Fig. 1.10) and after 1990 (Fig. 1.11). The majority of participants who studied empirical research related to the planetarium are middle and secondary school students (n = 15) and elementary school age students (n = 14) during the analog era of the planetarium. Other studies used college students (n = 7), multi-aged groups (n = 3), pre-service teachers (n = 1), planetarium directors (n = 1), and

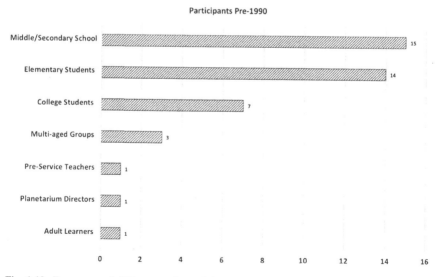

Fig. 1.10 Frequency of different study-participants in planetarium education research reports prior to 1990

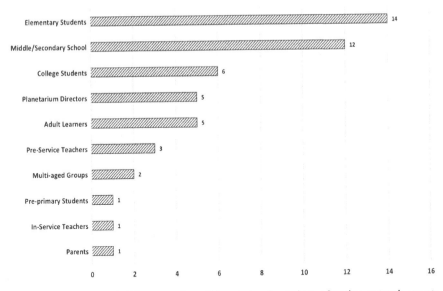

Fig. 1.11 Frequency of different study-participants in planetarium education research reports since 1990

adult learners (n = 1) as participants. Elementary school and secondary school students, elementary (n = 14) and secondary school age student studies (n = 12), are still the main groups of participants studied in empirical research related to the planetarium during the more recent digital era of the planetarium. Other studies used college students (n = 6), planetarium directors (n = 5), adult learners (n = 5), pre-service teachers (n = 3), multi-aged groups (n = 2), pre-primary students (n = 1), in-service teachers (n = 1), and parents (n = 1) as participants of interest during the digital age of the planetarium, which is much more diverse in the types of groups studied than in the analog era (Fig. 1.12).

Before the digital era, studies in the planetarium were mostly focused on evaluating students' understanding of how well they understand ideas covering multiple concepts (n = 11). More specifically, celestial motion (n = 8), stars (n = 5), and the Sun-Earth-Moon system (n = 4) were some of the main content areas of interest for researchers. As shown in Fig. 1.13, planetarium education researchers since mostly focus on students' understanding of the Sun-Earth-Moon system (n = 11) or general astronomy content that encompasses multiple topics (n = 10). A large portion of these studies also focus on celestial motion (n = 7), and a select few other studies focus on the moon illusion (n = 2) and cosmological concepts (n = 2). Note that if only one study covered any other content it was not included in the graph. For both dissertation and peer-reviewed studies, they reflect the same graphical shape for content incorporated in studies both for pre- and post-1990.

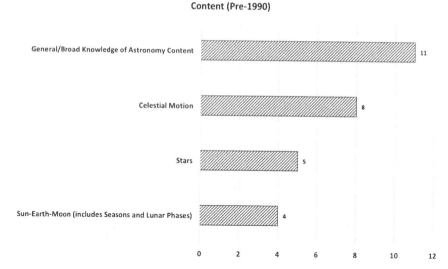

Fig. 1.12 Frequency of scientific content studied in planetarium education research prior to 1990

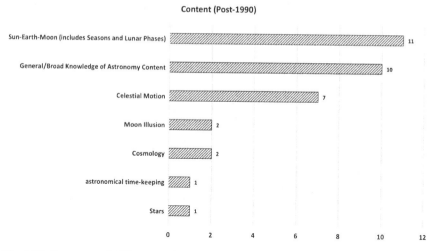

Fig. 1.13 Frequency of different study-participants in planetarium education research reports since 1990

The environmental setting where planetarium education research was conducted changed slightly before and after 1990. In all cases (Figs. 1.14 and 1.15), we see that most planetarium education research is done specifically for formal learning contexts in terms of developing understandings of how planetarium facilities were helping schools meet their school district specified learning targets. Before 1990, the overwhelming majority of studies were done under a formal context (n = 37),

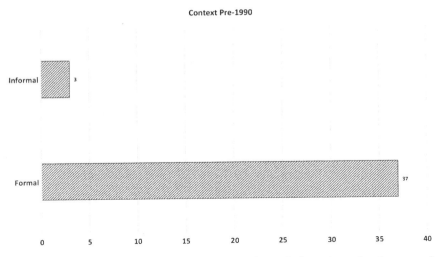

Fig. 1.14 Distribution of formal versus informal settings of planetarium education research reports prior to 1990

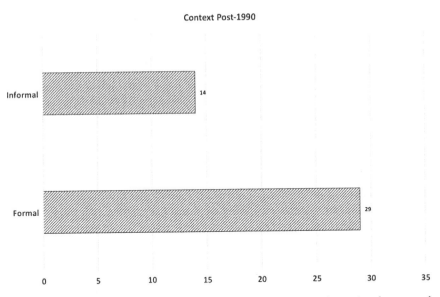

Fig. 1.15 Distribution of formal versus informal settings of planetarium education research reports since 1990

while only three studies were done in an informal context (Fig. 1.14). After 1990, the same pattern holds true but less drastically. Informal studies (n = 14) are still less common than studies held in a formal context (n = 29) (Fig. 1.16). It is likely that this is due to the fact that conducting studies on a controlled group, such as a

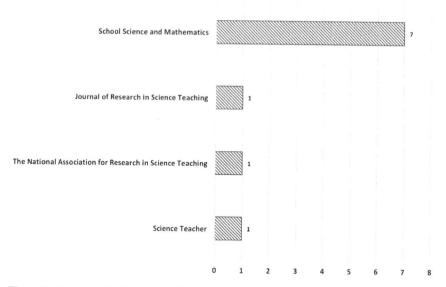

Fig. 1.16 Journals publishing planetarium education research since 1990

school group visiting a planetarium, is much easier than in a natural informal environment where addressing learning gains is much more difficult to assess.

Similarly, there are dramatic changes where planetarium education research was published before and after 1990. As shown in Fig. 1.17, the principle academic journal to publish the most peer-reviewed empirical works before 1990 was *School Science and Mathematics* (n = 7). There were only three other publications publishing peer-reviewed articles prior to 1990: *Journal of Research in Science Teaching*, JRST's one-time competitor, the *National Association for Research in Science Teaching*, and *The Science Teacher*. During the more recent digital era since 1990, the major journals publishing empirical planetarium education research studies include the *Journal of Astronomy & Earth Sciences Education* (n = 4), the *Planetarian* (n = 3), the *Journal of Science Education and Technology* (n = 3), the *Journal of Research in Science Teaching* (n = 3), *Science Education* (n = 2), and the now no longer publishing *Astronomy Education Review* (n = 2). Another 11 journals have published at least one empirical study on planetarium research education (Fig. 1.17).

Table 1.3 represents the most prominent characteristics defining each era. Each of these categories describes the shift in planetarium education research from the analog to the digital era. Planetarium education research was primarily conducted through dissertations and theses using quantitative methodologies, whereas today the majority of research uses both qualitative and quantitative methodologies and is published through the peer-reviewed process. Although fewer peer-reviewed articles were published pre-1990, George Reed at West Chester State College is

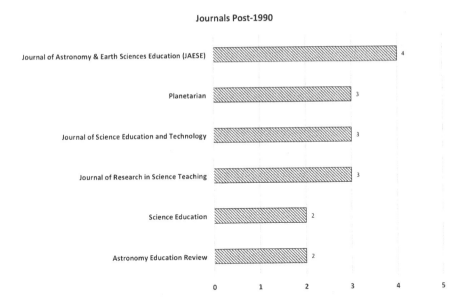

Fig. 1.17 Journals publishing planetarium education research since 1990

Table 1.3 Defining characteristics of pre- and post-1990 eras of planetarium education research

	Analog era (pre-1990)	Digital era (post-1990)
Most common document source	Dissertations	Peer-reviewed journal articles
Most common research method	Quantitative only	Both quantitative and qualitative/interpretive
Most commonly studied participants	Middle and secondary school students	Elementary Students
Most common scientific content	General astronomy content	Sun-Earth-Moon System
Most common environmental context	Formal education	Formal education
Most prolific journal	*School Science and Mathematics*	*Journal of Astronomy & Earth Sciences Education*

responsible for the majority of those works, and today Julia Plummer has published the bulk of studies regarding the planetarium while she was faculty at Arcadia University. Planetarium education researchers from both the analog and digital eras were mostly focused on K-12 students' understanding of astronomy content; it wasn't until digital planetariums that researchers narrowed their scope to focus specifically on the sun-earth-moon system, such as lunar phases and seasons. Pre-1990, the journal *School Science and Mathematics* was the most central platform for peer-reviewed, empirical planetarium education studies, whereas today the *Journal of Astronomy & Earth Sciences Education* has taken its place.

Taken together, the nearly century-long history of the planetarium education field shows a profound difference before and after 1990. Since 1990, the planetarium education field has responded to momentum changes in both the US educational system and an unprecedented availability in off-the-shelf technology. Whereas the pre-1990s golden age of planetariums could assuredly be characterized as unrestrained growth and enthusiastic construction of new facilities, there are few new planetarium facilities being constructed in the US. Indeed, most of the planetarium building efforts are actually in the remodeling and updating of pre-existing facilities, not in the construction of new facilities. This is in contrast to countries other than the US where there is a flurry of new planetarium facility construction. The changes to the planetarium education field, to name just a few, are: a dramatic shift from projecting the stars from internally illuminated star balls to computer-driven projectors; changing from live lectures delivered by an astronomy education expert to pre-recorded, multi-media productions; and a sea change in scientific content from shows designed to point out the constellations and planets moving in the current sky to delivering current news about scientific research advances concerning mysterious astronomical objects observed with high-technology equipment; and large theaters experiencing unexpected financial stress because of the faltering US economy. This financial issue concerns the increasing use of planetarium domes for high-revenue laser shows and grand-scale IMAX film screenings instead of the staple night-time star tours that garner sufficient revenue to make the financial books balance (viz., Reed 1994; Hitt 1999). These are illustrated in Table 1.4.

At the same time, the nature of planetarium educational research has been evolving recently. Perhaps most influential, the once locally-controlled K-12 school district curriculum is becoming more influenced at the federal level. Around 1990, several influential groups started aggressively advocating for new national consensus guidelines, including national standards and curriculum frameworks, to designate how much astronomy should be taught in all K-12 schools (Slater 2000).

Table 1.4 Characteristics of golden and modern planetarium education ages

Golden age of planetarium education	Modern age of planetarium education
Mechanical star balls and 35 mm slide projectors	Computer-driven video projection
Rapid construction of planetariums	Portable planetariums contribute to portfolio
Concentric seating around star ball	Unidirectional seating
Live-lectures	Turn-key multi-media presentations
Focus on moving celestial sphere	Focus on current scientific advances
Astronomy presentations	Music laser shows and IMAX movies
Locally controlled school district curriculum	National, common core curriculum standards
Most students are native english speakers	Rapidly growing diversity of student demographics

Moreover, a significant emerging challenge to the US' contemporary planetarium education community has been how to deal with rapidly growing diversity among K-12 students, many of which are quite different demographically than planetarium educators themselves. Although there are a wide variety of ways to classify which students are part of a minority is rapidly growing across the country. For example, public school children who are non-Native English speakers, also known as English Language Learners (ELLs), are a minority. During the 2012–2013 academic year, for instance, Alaska, California, Colorado, Nevada, New Mexico, Texas, and the District of Columbia had at least 10% of their students classified as ELL, with California having 22.8% of students being ELL (NCES 2015).

Taken together, there are tremendous forces shaping the way planetariums serve as a critical component of the larger US education portfolio. As a result, the future research questions pursued by planetarium education researchers will evolve as well. Future research endeavors are much more likely to focus on the roles of scientific visualization in helping non-Native English speakers engage in learning astronomy. Additionally, the role of the planetarium in shaping attitudes about national space policy and national scientific research priorities will need to be explored, as the national policy stakes are arguably higher today than they were even in the immediate post-Sputnik era corresponding to the 1958 US National Defense Education Act. Moreover, planetarium education researchers will have to engage in shedding the one-size fits all education approach and find innovative ways to individualize the planetarium learning experience. This will help people from different cultural backgrounds make sense of the world in their various processes. Finally, in an era of increased accountability for schools, the true academic value of student achievement in the planetarium will need to be revisited with new tools, community-agreed upon survey instruments, and more comprehensive research techniques available to planetarium education researchers.

References

Bishop, J. E. (1976). Planetarium methods based on the research of Jean Piaget. *Science and Children, 13*(8), 5–8.

Brazell, B. D. (2009). Planetarium instructional efficacy: A research synthesis. Doctoral Dissertation, Texas A&M University.

Dean, N. J., & Lauck, G. M. (1972). Planetarium instruction–using an open-sky test. *Science Teacher, 39*(5), 54–55.

Ehrenberg, R. G., Zuckerman, H., Groen, J. A., & Brucker, S. M. (2009). *Educating scholars: Doctoral education in the humanities.* Princeton University Press.

Erickson, F. (1985). Qualitative methods in research on teaching. In M. Wittrock (Ed.), *Handbook of research on teachsing* (3rd ed., pp. 119–161). NY: Macmillan.

Fletcher, J. K. (1980). Traditional planetarium programming versus participatory planetarium programming. *School Science and Mathematics, 80*(3), 227–232.

Gagné, N. (1989). The paradigm wars and their aftermath: A "historical" sketch of research on teaching since 1989. *Teachers College Record, 91*(2), 135–150.

Hitt, R. J.,Jr., (1999). *A national survey of planetarium directors operating public-school-owned planetaria*. Doctoral Dissertation, Virginia Polytechnic Institute and State University.

Hoff, D. B. (1970). A comparison of a directed laboratory versus an enquiry laboratory versus a non-laboratory approach to general education college astronomy. *Doctoral Dissertation*, University of Iowa.

Hunt, J. L. (1993). *Visual design in the planetarium*. Art, Science & Literacy: Selected Readings from the 24th Annual Conference of the International Visual Literacy Association. Pittsburgh, Pennsylvania.

Lipsey, M., & Wilson, D. (2001). Practical meta-analysis. *Applied Social Research Methods Series, 49*. (Thousand Oaks, CA: Sage).

Mallon, G. L., & Bruce, M. H. (1982). Student achievement and attitudes in astronomy: An experimental comparison of two planetarium programs. *Journal of Research in Science Teaching, 19*(1), 53–61.

"Planetariums of the World." www.ips-planetarium.org. International Planetarium Society, 28 August 2015. Web: 5 May 2016.

NCES. (2015). *the condition of education 2015 (NCES 2015-144), english language learners*. Washington, D.C.: U.S. Department of Education, National Center for Education Statistics.

Petersen, C. C. (2003). The birth and evolution of the planetarium. In *Information handling in astronomy-historical vistas* (pp. 233–247). Springer: Netherlands.

Plummer, J. D. (2006). *Students' development of astronomy concepts across Time*. Doctoral Dissertation, The University of Michigan.

Plummer, J. D., Schmoll, S., Yu, K. C., & Ghent, C. (2015). A guide to conducting educational research in the planetarium. *Planetarian, 44*(2), 8–24.

Reed, G. (1970). Is the planetarium a more effective teaching device than the combination of the classroom chalkboard and celestial globe? *School Science and Mathematics, 70*(6), 487–492.

Reed, G. (1973). The planetarium versus the classroom—an inquiry into earlier implications. *School Science and Mathematics, 73*(7), 553–555.

Reed, G., & Campbell, J. R. (1972). A comparison of the effectiveness of the planetarium and the classroom chalkboard and celestial globe in the teaching of specific astronomical concepts. *School Science and Mathematics, 72*(5), 368–374.

Ridky, R. W. (1975). The mystique effect of the planetarium. *School Science and Mathematics, 75*(6), 505–508.

Reed, G. (1989). *Dark sky legacy: Astronomy's impact on the history of culture*. Prometheus Books.

Reed, G. (1994). Who in the hell needs a planetarium? *Planetarian, 23*(1), 18–20.

Riordan, R. (1991). Planetarium education: A review of the literature. *The Planetarian, 20*(3), 18–25.

Rosemergy, J. C. (1967). An experimental study of the effectiveness of a planetarium in teaching selected astronomical phenomena to sixth-grade children. *Doctoral Dissertation*, University of Michigan.

Schatz, D., & Lawson, A. E. (1976). Effective astronomy teaching: intellectual development and its implications. *Mercury, 5*(4), 6–13.

Slater, T. F. (2000). K-12 astronomy benchmarks from project 2061. *The Physics Teacher, 38*(9), 538–540.

Slater, T. F. (2015). *Is the best astronomy education research 'grey'?* Retrieved from the *AstroLrner* blog at: https://astronomyfacultylounge.wordpress.com/2015/09/24/is-the-best-astronomy-education-research-grey/

Slater, S. J., Slater, T. F., Heyer, I., & Bailey, J. M. (2015). Conducting *Astronomy Education Research: An Astronomers' Guide, (2nd Ed.)*. Hilo, Hawai'i: Pono Publishing.

Slater, S. J.,Tatge, C. B., Schleigh, S. P., Slater, T. F., Bretones, P. S., McKinnon, D., & Heyer, I. (2016). iSTAR First light: Characterizing astronomy education research dissertations in the iSTAR database. *Journal of Astronomy & Earth Sciences Education, 3*(2), 125–140.

Smith, B. A. (1966). An experimental comparison of two techniques (planetarium lecture-demonstration and classroom lecture-demonstration) of teaching selected astronomical concepts to sixth grade students. *Ed.D. Dissertation*, Arizona State University.

Smith, T. V. (1974). *The planetarium in education: A review of the literature.* Nova University. Retrieved from ERIC at http://files.eric.ed.gov/fulltext/ED111658.pdf

Sneider, C. I., Eason, L. P., & Friedman, A. J. (1979). Summative evaluations of participatory science exhibit. *Science Education, 63*(1), 25–36.

Sonntag, M. (1988). Focus on education spatial ability, science, and sex. *Planetarian, 17*(4), 37–41.

Sonntag, M. S. (1989). The effect of a combination of planetarium and laboratory instruction in the improvement of student spatial ability. *Planetarian, 18*(3), 47–49.

Soroka, J. J. (1967, October). The planetarium and science education. *Projector-Publication of the Great Lakes Planetarium Association,* 12–23.

Sunal, D. W. (1973). *The Planetarium in Education: An Experimental Study of the Attainment of Perceived Goals.* Doctoral Dissertation. University of Michigan.

Sunal, D. W. (1976). Analysis of research on the educational uses of a planetarium. *Journal of Research in Science Teaching, 13*(4), 345–349.

Tuttle, D. E. (1965). *Effects of the use of the planetarium upon the development of spatial concepts among selected sixth grade students in Elgin* (Master's thesis), Northern Illinois University.

Tuttle, D. E. (1967, October). Elgin's research in planetarium curriculum. *Projector-Publication of the Great Lakes Planetarium Association,* 9–12.

Warneking, G. E. (1970). Planetarium education in the 1970's—Time for assessment. *The Science Teacher, 37*(7), 14–15.

Wright, D. L. C. (1968). Effectiveness of the planetarium and different methods of its utilization in teaching astronomy. *Doctoral Dissertation*, University of Nebraska.

Wyatt, R. (2005). Planetarium paradigm shift. *Planetarian, 34*(3), 15–19.

Chapter 2
Overview of Planetarium Education Research Methods

The modern digital planetarium, with its inherent ability to show the universe as a real-time three-dimensional model, as opposed to the old-style, earth-centered view of the optical star projector, provides unique opportunities for a fresh look at education research in the planetarium. Now reaching a level of maturity in the second decade of its use, where digital tools are becoming commonplace, the digital planetarium offers a chance to ask old questions with a new perspective that has real datasets of the entire known map of the universe.

Can students identify constellations in the night sky better if they learn in a planetarium or in the classroom? Can students better predict planetary retrograde loops if they have watched planets move using spatially different viewpoints in a digital planetarium? Which astronomy topics are best taught in the planetarium? How much do students enjoy learning in the planetarium? Do classes held in a planetarium improve attention and retention? Do visits to the planetarium increase the number of students entering the science and engineering pipeline? How do students make sense of the celestial sphere? These, and countless other questions, dominate the realm of planetarium education research.

Sometimes asking these questions results in refereed journal publications or in doctoral dissertations that add to the academic literature base stored in university libraries. Other times, asking these questions results in evidence-based talks given at professional planetarium society conferences that consequently urge commercial planetarium program developers to dramatically change the way that information is presented to audiences. These developers come up with new user friendly tools, which make learning in the planetarium more durable and effective. And, sometimes, like in much of scientific inquiry, study results end up creating more questions than actually providing solid actionable answers. This chapter provides a brief introduction to the major methods and underlying foundations adopted by the broad community of scholarly planetarium education researchers in their academic pursuits.

© The Author(s) 2017
T.F. Slater and C.B. Tatge, *Research on Teaching Astronomy in the Planetarium*,
SpringerBriefs in Astronomy, DOI 10.1007/978-3-319-57202-4_2

Planetarium education research typically falls under the broader scientific domain of discipline-based education research. Discipline-based education research, as defined by Slater (2015a, b) and colleagues, is the scientific enterprise of making systematic observations and conducting ethical experiments of learners and their environments in order to develop predictive and explanatory models of teachers, teaching, learners, and learning, in the service of enhancing learning. Such a definition is consistent with other authors, including those representing the National Research Council and the National Academy of Sciences (Singer et al. 2012). Although some traditional scientists occasionally mock the science education research enterprise as being subjective and lacking the objective rigor of conventional bench-based laboratory science as a whole, a systematic and data-driving scientific approach to better understanding the process of teaching and learning is a respected and formally supported part of the larger scientific community. The largest professional scientific societies have all issued position statements emphasizing the value of carefully conducted scientific research in the domains of teaching and learning. Figure 2.1 lists some of these position statements.

There are a wide variety of domains in which a planetarium education researcher might apply their scientific inclinations to better understand the role planetariums play in the educational landscape. One could study how people learn from the planetarium, for certain. And, one might also choose to study how people feel after a planetarium learning experience. Additionally, one might choose to compare the type of planetariums in formal school-building environments to those in museum and science center environments. Alternatively, one might choose to compare the nature and specific scientific content in the programming of permanent brick-and-mortar planetarium buildings versus the inflatable portable planetariums. One might even choose to study the range and domain of the academic qualifications of individuals working in the planetarium field. What about studies on learning via computer desktop planetarium simulation programs on a flat screen monitor? Given the seemingly endless range of possibilities, it seems necessary, albeit unappetizing, to constrain planetarium education research to some degree in order to begin making sense of it.

One of the most straightforward approaches to constraining the range of planetarium education research efforts might be to separate the vast number of study possibilities such that studies fit into one of two categories: descriptive and experimental. On one hand, descriptive studies provide an objective, third-party narrative of what happened during an educational event, the sequence of how it happened, and why the event took place. Such studies can provide long descriptions or conduct a comparative analysis of different educational situations. An example would be to describe the number and ages of people who visit a planetarium. Another example would be to survey planetarium directors across the country and tabulate which show schedule times attract the largest number of attendees.

On the other hand, experimental education studies attempt to measure a change and attribute that change to an educational intervention. There are at least two traditional ways to conceptualize a basic experimental study. One way is to take a group of study-participants and measure something about them—their knowledge

American Astronomical Society Position Statement on Research in Astronomy Education
Adopted 2 June 2002
https://aas.org/governance/council-resolutions#edresearch

In recent years, astronomy education research has begun to emerge as a research area within some astronomy and physics/astronomy departments. This type of research is pursued at several North American universities, it has attracted funding from major governmental agencies, it is both objective and experimental, it is developing publication and dissemination mechanisms, and researchers trained in this area are being recruited by North American colleges and universities.

Astronomy education research can and should be subject to the same criteria for evaluation (papers published, grants, etc.) as research in other fields of astronomy. The findings of astronomy education research and the scholarship of teaching, when properly implemented and supported, will improve pedagogical techniques and the evaluation of both teaching and student teaching.

The AAS applauds and supports the acceptance and utilization by astronomy departments of research in astronomy education. The successful adaptation of astronomy education research to improving teaching and learning in astronomy departments requires close contact between astronomy education researchers, education researchers in other disciplines, and teachers who are primarily research scientists. The AAS recognizes that the success and utility of astronomy education research is greatly enhanced when it is centered in an astronomy or physics/astronomy department.

American Physical Society Policy Statement on Research in Physics Education
Adopted May 21, 1999
https://www.aps.org/policy/statements/99_2.cfm

In recent years, physics education research has emerged as a topic of research within physics departments. This type of research is pursued in physics departments at several leading graduate and research institutions, it has attracted funding from major governmental agencies, it is both objective and experimental, it is developing and has developed publication and dissemination mechanisms, and Ph.D. students trained in the area are recruited to establish new programs. Physics education research can and should be subject to the same criteria for evaluation (papers published, grants, etc.) as research in other fields of physics. The outcome of this research will improve the methodology of teaching and teaching evaluation.

The APS applauds and supports the acceptance in physics departments of research in physics education. Much of the work done in this field is very specific to the teaching of physics and deals with the unique needs and demands of particular physics courses and the appropriate use of technology in those courses. The successful adaptation of physics education research to improve the state of teaching in any physics department requires close contact between the physics education researchers and the more traditional researchers who are also teachers. The APS recognizes that the success and usefulness of physics education research is greatly enhanced by its presence in the physics department.

Fig. 2.1 Examples of professional scientific society statements on education research

Single Group Pre-test-Posttest Experimental Design				
Pre-test	→	Educational Intervention	→	Post-test

Fig. 2.2 Schematic of single group pre-test-posttest experimental design

or their attitudes for example—before and after an educational intervention, and determine how much change has occurred (Fig. 2.2). Another way is to take a sample of study subjects and randomly assign them to different treatments and measure how groups differ as a result of their various assigned treatments (Fig. 2.3). It is important to emphasize that study subjects must be randomly assigned to various treatment groups to be a true experiment. The reality of working with school children, for instance, is that it is incredibly difficult to randomly assign students to different treatments. Instead, what usually happens for pragmatic reasons of scheduling and supervision is that students are not randomly assigned to different treatments. This results in a study being classified not as experimental, but as quasi-experimental. Being classified as a quasi-experimental research study is not necessarily a pejorative label; rather, the label is an accurate description of how a study was done as a result of how participants are selected and assigned treatments.

Unceremoniously dividing the breadth of planetarium education research work into just two broad categories is probably insufficient to capture the breadth and depth of the field. An early scholarly effort to constrain what constitutes the domain of education research was advocated by Novak (1963). Novak (1963) felt strongly that science education research is best done by people who are inclined toward, and perhaps formally trained in, scientific observation and experimentation. Underlying his thinking was that educators of the day were naturally focused on nurturing youth as growing and evolving humans and would be largely unable to make rationally objective and evidence-based conclusions about comparative science education interventions. More than who should be doing science education research, he also argued passionately that the role of science education research should be entirely focused on advancing a priori explanatory theory through experimentation. In other words, the goal of science education research should be to develop sufficiently complete educational theories through repeated experimentation so that one could successfully predict the educational outcome of a proposed educational intervention without actually doing the experiment.

Two Group, Post-test-Only Control-Group Experimental Design						
Randomization of Participants into Groups	Experimental Group	→	Educational Intervention	→	Post-test
	Control Group	→	No Intervention	→	Post-test

Fig. 2.3 Schematic of two group, post-test-only control-group design

Novak (1963) divided the landscape of science education research into four domains: baseline studies, correlation studies, experimental studies and curriculum studies. *Baseline studies* are those census surveys of the field that establish a snapshot of the current status and characteristics of planetarium education research that are normative descriptions of the field: How many planetariums are there? Who attends planetarium shows and what time of day? What is the portfolio of academic qualifications of planetarium lecturers? These studies are important for individuals who want to watch the evolution of planetarium education and understand the leverage points of engaging potential populations of new attendees.

A second category of studies Novak (1963) identified are *correlation studies*. Correlation studies are those in which two or more variables are compared to determine how often they occur together. An example question would be: do students have more achievement gains when learning in larger sized domes, as compared to smaller domes? Correlation studies are useful, but must be interpreted with a degree of caution. An often repeated mantra in planetarium education research is that correlation does not equal causation. In the example cited, larger domed facilities, such as those in museums and science centers, might be more often visited by students representing a distinct socio-economic group as compared to smaller domes. This might account for the differences in achievement gains seen in pre-testing and post-testing visiting students. In education research, a participant's socio-economic status can have a major influence on achievement and attitude changes as a result of an educational intervention. Such a scenario requires researchers to carefully describe the characteristics and demographics of any and all study participants.

The most powerful research studies that Novak (1963) identifies are scientific *experimental studies* (viz., Figs. 2.2 and 2.3). These are education research studies where all, or most, variables are identified and only one independent variable is carefully manipulated at a time. All other variables are held constant, or randomized, so that a responding, dependent variable can be measured. In this case, the responding, dependent variable can be reasonably attributed to changes in the manipulated, independent variable. For example, if one group of students sees a planetarium show that includes significant humor, and another group of students sees the same show without humor included, one can reasonably attribute any observed differences in learning gains to the inclusion or exclusion of included humor if the initial equivalency of the two groups can be firmly established (e.g., Fisher 1997). In this illustrative example, as in many others, the demographics of any participants in the study need to be carefully described. Changes in achievement due to the inclusion or exclusion of humor in a planetarium program may be the result of differences in cultural background, so this needs to be accounted for. Furthermore, each population of students may provide varying results because of differences in cultural backgrounds.

A final category of studies that does not fit into the three previously identified by Novak (1963) is what he terms *curriculum studies*. This category includes studies on the astronomical topics we teach in each facility and at each grade level. For example, in the US, one might be interested in describing how the *Next Generation*

Science Standards are being manifested in planetarium education (Schleigh et al. 2015). One of the problems with conducting this sort of academic work is that authors identify what is being taught in education systems by popularity due to high frequency, rather than what is established as core ideas of a discipline. Planetarium education systems in the US have struggled with this over recent decades, in particular, because of generous funding from NASA. These funds to education and outreach programs enhance the public's support of particular NASA science and specific NASA missions, rather than astronomy and space science as a coherent whole. Novak (1963) argued that curriculum studies should only be taken on by individuals with an exceptionally strong background and deep understanding of the philosophy of science and the historical development of scientific disciplines. More critically, Novak (1963) argued that curriculum studies are most often unworthy of being reported in the archival scholarly literature as they are generally focused at the local level where the study was conducted. They also usually have little or no value to the larger science education community, unless some promising new research method is utilized to conduct the curriculum study.

Ball and Forzani (2007) advocate a vision of education research that offers guidance in terms of the type of studies the planetarium education research community could most reliably include as their domain of education research. In response to the value of education research studies too often being questioned by skeptics, well-respected, education thought leaders Ball and Forzani (2007) argue that the education researcher—and in the current case, the planetarium education researcher—would be best served by focusing their scholarly studies on research within education, instead of studying phenomena that is only related to education. In this sense, examples of study on topics related to education would be counting and classifying the number of planetariums by dome size or the reasons for people returning to see each and every planetarium program a facility presents. These are not key aspects to understanding the underlying mechanisms of planetarium education.

Ball and Forzani (2007) instead argue that the planetarium education research that really matters is that which focuses on what they call *instructional dynamics* and *educational interactions*. In this sense, they encourage the community's scholarly researchers to concentrate on how planetarium programming can uniquely interpret and present astronomy to students. They also support research on the specific mechanisms that drive students' responses to that stimulus either in terms of achievement or attitude. An example consistent with this perspective in planetarium education research would be which specific learning interventions occurring in the planetarium cause changes in students' understanding—be they hands-on with manipulates or minds-on with cognitively challenging mental engagement tasks. These are characterized by studying interactions between the planetarium stimulus and students' responses. Some examples that are inconsistent with this modern idea of worthy research are studies that comprehensively investigate students' attitudes or beliefs about astronomy or career aspirations, but provide no insight on the educational transactions at planetariums that influence students' thinking. In other words, planetarium education research that identifies changes in students, but is unable to robustly establish causal mechanisms, falls far short of Ball and Forzani's (2007) vision.

Ethical Guidelines for Education Research

What was judged appropriate planetarium education research fifty years ago, and what will be considered the most useful planetarium education research going forward, is evolving substantially. One of the monumental changes occurring in planetarium education research over the last two decades is an enthusiastic national focus on researchers being highly ethical in human studies. In the 1960s and 1970s, seminal planetarium education research studies were done by surveying students upon entering and exiting the planetarium. The question about whether or not these early era study participants were willing subjects was not really a concern for most planetarium education researchers. In those days, students were relatively compliant and when an authority figure in education, such as a teacher or a planetarium lecturer, asked students to voluntarily complete a survey without compensation, students most often acquiesced to the request without complaint from either themselves or their parents. This situation was not unique to planetariums, but was characteristic of individuals in most educational systems. In essence, when students would enter the planetarium, or classroom, it seemed perfectly appropriate to give them a survey about their understanding of, or attitudes toward, astronomy and expect them to complete the survey.

A perspective that data could be collected from students without their well-informed understanding and consent has changed dramatically in the last two decades. Today, planetarium education researchers are bound by ethical guidelines that require human subjects to be fully informed of any study they are participating in, no matter how insignificant it might appear to the researcher (Slater et al. 2015a). Normally, this is done by verbally describing the study and its potential benefits and risks. No study has zero risks. A participant could unexpectedly become anxious or distraught while completing a survey if they do not know the answers. More importantly, no study has absolutely zero costs. If you require a participant to complete a five-minute survey, that is five-minutes of their personal disposable time that they could have used in another way of their own choosing, perhaps responding to a critical email instead of completing a survey. Also, asking participants to complete a survey in the planetarium could take away time that could have been spent actually learning something from the planetarium. In other words, there are risks and costs to being a research participant that novice planetarium education researchers might not realize. Most often, voluntary participation is solicited and consented to when researchers provide both an orally spoken description and a written consent form to be signed by the participant. An illustrative example is given in Fig. 2.4. Parents or guardians may also need to sign a consent form if study participants are not of legal age (18 in most states).

My name is _____and I want to tell you about a research study I am doing. A research study is usually done to find a better way to understand how things work. In this study, I want to find out more about the best way for students to learn _____.

The research study you will participate in will take place _____. During the research study you will _____. The lesson and assessments should take no longer than _____.

There is minimal risk involved in the participation of this research study. There is a chance you may feel anxiety before, during, or after the pre- and post-assessments, or you may become embarrassed during the lesson if you do not know the answer to questions I ask. But these risks are minimal and any embarrassment, anxiety, or discomfort that this research may cause is no greater than that of any other risk that you may encounter in a classroom, during a lesson, or while taking a test. If you do feel anxiety or discomfort at any time just let me or your teacher know and we will allow you to take a small break or exit the room.

There are benefits for you to participate in this research study. The main benefit is an increased knowledge on _____, which is part of your curriculum. The knowledge gained during this research study will also be helpful for your in-class and standardized statewide tests. This study also has the potential to help teachers and curriculum developers to better teach future students. All of your data will be kept confidential. Each of you will be identified using a numerical code, which will be given to you during the initial meeting. You will use this numerical code on all assessments during the study. No identifying characteristics will be used in the presentation of the data. If you are referred to in any class assignment or publication, only your pseudonym will be used.

I will not discuss your name with anyone other than your teacher or my colleagues. All data, including signed consent and assent forms, master lists with your name and corresponding numerical code, pre- and post-test scores, and spatial ability scores will be stored in a digital format on a password protected hard drive that only myself and my colleagues can access. All hard copies will be stored in a locked filing cabinet in my home office that only I will have a key to. The research summary, all tests, signed consent forms, and signed assent forms will be stored for three years after the research is completed, until November 30, 2019. Your data may be used for future class projects I am assigned or future publications, but again, your identification will be confidential and no names will be provided. Your participation is voluntary and your refusal to participate will not involve penalty or loss of benefits to which you are otherwise entitled, and you may discontinue participation at any time without penalty or loss of benefits to which you are otherwise entitled.

Also, you understand that your refusal to participate or your withdrawal at any point will not affect your course grade or class standing.

If you choose not to participate in this study at any time please let your teacher or myself know. You will still be expected to participate in the lessons as part of your science curriculum but no data from your surveys will be collected or recorded. I am asking to use this data for research purposes.
If you have any questions or concerns about participating in this research study please contact myself_____ or my supervising colleagues at _____.

_____   _____
PRINT NAME   SIGNATURE

Fig. 2.4 Illustrative example of a human subjects consent signature form for minors

There are times and situations, however, when a particular study design benefits from a scenario in which participants can be deceived briefly about the nature of a study by the researcher, but such situations are nuanced, delicate, and rarely are part of the planetarium education research world. An oversimplified example of a rapidly conceived research study on the role of narration during a planetarium program might be where English–only speaking planetarium visitors are deliberately misled and misinformed that the only narrated soundtrack available is in Spanish, and that they will have to learn astronomy as best they can in this less than advantageous situation. The bottom line is that participants can only consent to be included in a research project when they are fully informed about the potential benefits and risks.

The types of experimental studies briefly mentioned earlier and illustrated in Figs. 2.2 and 2.3, can vary considerably in terms of specific configurations. However, nearly all research designs in planetarium education research surround the same theme: what was the impact of visiting the planetarium? The five most common study designs to uncover what happens as a result of learning in the planetarium are illustrated in Fig. 2.5.

The most basic study design is the snapshot case study. In this study design, planetarium attendees are surveyed after they have visited the planetarium. The advantage of this particular design is that it is relatively easy to administer—often in the form of an exit survey. The disadvantage is that the researcher does not know what the participants' responses were before they attended the planetarium and, as such, cannot robustly know that the planetarium experience itself had any impact at all. Instead, a stronger research design is the single-group, pre-test—post-test design. In this case, a researcher is able to establish what the incoming responses

Fig. 2.5 Illustrations of common study designs

R = Random Placement of Participants into Groups;
EI = Educational Intervention;
O/M = Observation/Measurement

ONE: Snap-Shot Case Study
    __   EI   O/M

TWO: Single-Group Pre-test-Post-test Design
O/M   EI   O/M

THREE: Pre-test-Post-test Control Group Design
R   O/M   EI   O/M
R   O/M   __   O/M

FOUR: Post-test-Only Control Group Design
R   __   EI   O/M
R   __   __   O/M

FIVE: Solomon Four-Group Design
R   O/M   EI   O/M
R   O/M   __   O/M
R   __   EI   O/M
R   __   __   O/M

are from a pre-test and any differences in responses on a post-test can be roughly attributed to the planetarium learning experience itself. The disadvantage to this design is that one cannot know the impact of taking the pre-test. For example, it is possible that taking the pre-test unsuspectingly telegraphs or cues participants to pay more attention to some parts of the planetarium learning experience than they otherwise would have. In this case, the experience of taking the pre-test itself becomes an uncontrolled, confounding variable.

Control Group Design

To mitigate for the potential impact of taking a pre-test on the study participants' post-test responses, most planetarium education researchers adopt the pre-test—post-test, control group design. This study design is characterized by randomly dividing the study population into two groups, where one receives the educational intervention, and the other does not. In addition to mitigating for any unknown impacts of taking the pre-test on study participants, this design also allows the researcher to establish an initial equivalence among the two groups. If the two groups are not initially equivalent, results could be unknowingly influenced by one group that is already educated in the planetarium or previously had their attitudes positively enhanced. This research design has the further advantage of being able to more confidently attribute differences in post-test scores to a specific educational intervention in the planetarium compared to the two previously mentioned designs.

A particular characteristic of the pre-test—post-test control group design is worth noting, and it is related to the term *control group*. Planetarium education researchers often describe two comparison groups of study participants as either being members of the *experimental group* or the *control group*. Most often, the study participants that were described as the experimental group was the group that learned in the planetarium and the group of study participants that learned in the traditional classroom is called the control group. In modern writing, planetarium education researchers should be most appropriately referring to the two involved comparison groups as *treatment-1* and *treatment-2* groups. This is because in a traditional, scientific experimental research study, one of the groups should receive a treatment—in this case an educational intervention—and the other group should receive no educational intervention treatment at all, if it is actually a control group in the strictest sense. Very rarely in planetarium education research do we read about studies with a non-treatment control group that received no educational intervention whatsoever, even though this is how they are sometimes described. There is an ethical reason for this as well. Most planetarium educators do not want to academically disadvantage half of the students by not giving them any instruction in astronomy. As a result, researchers either give students an alternative learning experience or, what is now more commonly done, provide the non-planetarium learning group a chance to go to the planetarium after the experimental study has concluded.

Depending on the specific nature of the study, even the pre-test—post-test control group design does not fully account for the impact of participants taking a pre-test. The most well-known problem related to pre-tests has to do with studies involving tests of spatial reasoning. It has been observed that scores improve each time students take a spatial reasoning test, even when there is no educational intervention. In other words, simply taking a spatial reasoning test improves your score the next time you take a spatial reasoning test (Heyer et al. 2013). Somehow, the act of taking the test increases your ability to do spatial reasoning. In response, some researchers argue the advantages of using a post-test-only control group design over the pre-test—post-test control group design, depending on the precise nature of the research question.

Perhaps the strongest research design available to planetarium education researchers is known as the Solomon four-group design. Illustrated in Fig. 2.5, the Solomon four-group design contains two extra control groups, which allows the researcher to explicitly test the extent to which the pretest itself has an effect on study-participants' post-test responses. Admittedly, this powerful research design is much more complex and requires considerably more effort on the part of the planetarium education researcher to conduct. Campbell and Stanley (1963) convincingly argue that when the Solomon Four-Group Design is impractical, the post-test-only control group design is a stronger design than the pre-test—post-test control group design. It is more powerful to argue that two groups were initially equivalent than to argue that a pre-test had negligible impact on the participants prior to an educational intervention.

Validity and Reliability

In any study, understanding the nuanced strength and weaknesses of the measurement instruments themselves is usually the most challenging. In all cases, researchers must make a compelling argument that the instruments used are both reliable and valid. The concepts of reliability and validity can be likened to synonyms for precision and accuracy, respectively. Reliability refers to "being consistent in the sense that a subject will give the same response when asked again". That is to say, if one was to step on a scale multiple times and it gave results such as 135, 134, 134, and 135, one could say that that instrument was reliable. However, if one stepped on the scale and the results were much more varied (e.g., 128, 142, 136, and 112), then this would show that the instrument was unreliable and could not be counted on to give precise, consistent answers. Validity, on the other hand, has more to do with measuring "what is intended to [be] measured and accurately reflecting the concept". Extending the same example, if one knows that the weight should actually be 115 lbs but the scale is still reading 135, 134, 134, and 135 there must be some sort of systematic error.

In order to test for reliability of an assessment tool, there are a few different methods such as test-retest, equivalent-forms, internal consistency, and interscorer, all of which calculate a correlation coefficient to determine its reliability. The test-retest reliability method is computed by administering a test over time; equivalent-forms reliability uses multiple forms of a test created to measure the same construct; interscorer reliability is determined by having multiple scorers agree on the results given; and internal reliability refers to "how consistently the times on a test measure a single construct or concept". Most tests are supposed to measure a one-dimensional construct, and tests that do this have high internal reliability. However, other tests, such as the *Test Of Astronomy STandards* (TOAST), are multidimensional so in order to determine its internal reliability, each construct must be measured separately (Slater 2015). To calculate the internal reliability, researchers use the Chronbach alpha coefficient, most commonly called the "coefficient alpha" (Chronbach 1951). The coefficient alpha is calculated using the number of items on a test and the average correlation coefficient. A coefficient alpha higher than 0.7 is typically considered internally reliable and anything less than that could be due to test fatigue or some other underlying cause.

Validity can be a bit more difficult for which to argue. However, validity is just as important as reliability. As Chronbach says in his 1991 article, "a test may be excellent in other respects, but if it is wrongly interpreted it is worthless in that time and place". Ensuring validity is done by gathering evidence that supports the inferences and findings made by the researcher. Validation is one of the most important reasons a research study should be seated firmly within the context of the field and should "stand on the shoulders of giants". There are four main types of validation: construct, face, criterion, and content validity. Most validation starts out with face validity, by the researcher simply saying to his or herself, as well as asking an outsider's expert opinion, "it looks like it'll work". While it is important for the researcher to reflect on how valid their assessment tool seems, it is highly subjective and requires other forms of validation. Construct validity is determined by establishing the test measures the construct was designed to measure. For example, if an exam was trying to evaluate students' understanding of astronomy concepts, the developer should ensure that the reading level is well below the students' grade level to make sure reading ability does not affect the results. Criterion-related validity uses tests that have been previously validated for measuring a similar construct, to see if the results correlate with those outcomes. Lastly, content validity determines if the content used on the test is appropriated considering the subject area that is being assessed. For example, if a researcher wants to know how well students understand lunar phases, asking a question about celestial motion may be irrelevant. This type of validation should be done by asking experts in the field.

More specifically, a researcher may wish to determine the difficulty of certain items and evaluate which questions students "who get it" get correct versus questions students "who don't get it" get correct. These can be measured by calculating item difficulty and item discrimination. Item difficulty is a value given from 0 to 1 and equates to the percentage of students who answered an item correctly. If an item has a difficulty of 0.30 and a high discrimination score, then the students who did well on the overall exam are likely to make up the 30% of students who answered the question correctly. If the discrimination is low or negative for this item, it means that students who did not do well on the exam probably guessed on this item. Item discrimination can be determined by calculating the biserial-R. It is a correlation statistic from −0.1 to +1.0. If the score is near 0 it shows that answering that question "has nearly no relationship to overall matter of the material". Negative values indicate that students who did well on the overall exam did poorly on that particular item. These types of questions should be analyzed for flawed construction (Slater and Adams 2003).

When analyzing the quantitative and mixed methods studies submitted to the iSTAR database, we prefer articles that have clear and specific methods, as well as results sections explaining the procedures used to calculate whether results are reliable, validated, and significant. To ensure reliability, especially of a self-made assessment instrument for a specific study, we prefer to see a coefficient alpha of at least 0.7 or higher to ensure internal reliability. An additional reliability test would also help, such as the test-retest or interscorer methods. As for validation of a test instrument, numerous methods of validation are important. Construct validity and content validity ensures the researcher is measuring what they intend to measure and use appropriate content-related items. Criterion-related validity can be a more objective voice to the validation process, in order to determine the new assessment instrument's correlation with a previously deemed valid and reliable instrument. Face validity ensures that the overall structure and construction is sound. As reliability can be shown statistically, validation should be discussed in length in a study and it should be clear that the researchers went to exhaustive and extensive length to prove the validity of their instrument. Due to the complex nature of designing an instrument, it is typically best if researchers use a previously constructed instrument that has already been checked for reliability and validity. Hence the reason instruments such as the TOAST are so important for each field. Universally used instruments in a field also benefit the literature by providing a way for researchers to compare the outcomes of their studies. These instruments can also provide insight into the effectiveness a variety of dimensions that novices should be learning about in introductory astronomy courses based on different instructional methods, curriculum, and activities.

One example showing the importance of validity is a classic planetarium education research study done by Akey (1973). Akey (1973) conducted a pre-test—post-test study and found statistical significance in students' learning gains from using the planetarium on 56 astronomy concepts. However, reflecting on his assessment instrument today shows a lack in face validity. First, the test's reading level for the second grade students was too difficult. Also, 56 concepts would now be considered far too many astronomy concepts to teach such a young group of students in that short amount of time. Lastly, students were able to answer a large percentage of these answers correctly on the pre-test which is highly unlikely given what other research on planetarium education stated about the students' understanding of astronomy at that age level. Due to these validation issues, researchers now know that using the results from a study like this would be unscholarly and futile. The ability to reflect on the validation and reliability of previous studies is critical to understanding the quality of research that can truly help fuel the future of astronomy education research.

Even when the validity and reliability of an instrument has been thoroughly discussed, there are situations when these can be threatened by confounding issues beyond the survey instrument itself. Examples include when major newsworthy events occur between observations, when exhaustion (or maturation) occurs during or between observations, when observers or scorers standards drift from grading survey answers early in the study to late in the study, unknown and unintentional selection bias of selected participants perhaps from a particular socio-economic group, and last, but not least, interactive effects due to the mere presence of survey instrument or an observer in the room taking field-notes.

Bloom's Taxonomy

Almost all survey instruments, regardless of their form, attempt to quantify not just the amount of learning that occurs or the extent of attitude change, but the actual depth and quality of that learning. Most planetarium education researchers lean heavily on the longstanding theoretical work done decades earlier by Bloom (Anderson et al. 2001; Bloom 1956; Krathwohl et al. 1964). Bloom's (1956) initial efforts were based on the need to classify knowledge survey questions in terms of how many cognitive resources were required to answer them correctly. He started by classifying questions from a ranging of relatively easy recall questions and, with increasing cognitive complexity, to highly complex questions requiring the synthesis of numerous ideas in order to evaluate novel situations. Eventually, he concluded that assessment questions—and their corresponding learning objectives—could be categorized consistently into six cognitive levels. Known broadly as the *cognitive domain*, Bloom's six-level taxonomy ranges from the lowest cognitive levels to the highest cognitive levels in the sequence: (1) Knowledge; (2) Comprehension; (3) Application; (4) Analysis; (5) Synthesis; and finally, (6) Evaluation. In Fig. 2.6, each label refers to the cognitive tasks required to complete a task successfully. Note

Bloom's Taxonomy of Cognitive Domain Learning Objectives		
Levels	*Names*	*Characteristics*
Level One	Knowledge and memory	*Identify, list, repeat, recognize*
Level Two	Comprehension and understanding	*Locate, represent, categorize, explain*
Level Three	Application	*Relate, calculate, organize, use*
Level Four	Analysis	*Compare, contrast, experiment, discriminate, structure*
Level Five	Synthesis	*Compose, create, generalize, relate, construct*
Level Six	Evaluation	*Judge, estimate, criticize*

Fig. 2.6 Bloom's taxonomy of cognitive domain learning objectives

that each label has evolved somewhat over the decades. Successfully identifying the brightest star in the night sky as Sirius would be classified as a knowledge-level task, and evaluating a prediction of Earth's observed retrograde loop duration as seen from the surface of Mercury would be a synthesis-level task.

The precise language that planetarium education researchers use to describe Bloom's Taxonomy varies somewhat from one person to the next. While one researcher might refer to the knowledge level as the lowest level and the evaluation level as the highest level, another researcher might use words vaguely suggesting that the levels are inverted by ascribing the knowledge level as being the shallowest level and the evaluation level as being the deepest level, rather than the highest level. Be that as it may, Bloom's Taxonomy has been a highly valuable part of planetarium education researchers being able to describe the specific depth of understanding that they are testing and relating the impact that planetariums have on learning.

Many planetarium education researchers are also often interested in the evolving attitudes, values, interests, feelings, and motivations of people engaged in learning in the planetarium. Along with Krathwohl et al. (1964) and colleagues, Bloom also advanced a taxonomy describing the *affective domain*. A learner that has achieved affective learning at the lowest level of the affective domain is an individual who is clearly open to new experiences and willing to devote emotional energy to a person, environment, or event. In contrast, an individual who has achieved the highest level—or deepest level, according to some—is one who has internalized their feelings to the point that they will obviously behave in a predictable way, responding to their well-developed worldview. Bloom's definitions of the affective domain are not as widely adopted in the planetarium education research community as those descriptions of the cognitive domain. Nonetheless, Bloom's taxonomy of the affective domain is much more precise for describing attitudes rather than just the participants' high or low attitudes and levels of interest in a topic (Fig. 2.7).

Understanding the process of learning in a planetarium in terms of their specific depth of comprehension—cognitive domain—and characterizing their wealth of attitudes, values, and interests—affective domain—is indeed complex. Adding to this complexity is the environmental context where learning in the planetarium

Bloom's Taxonomy of Affective Domain Learning Objectives		
Levels	*Names*	*Characteristics*
Level One	Receive	*Open to experience, attend, acknowledge*
Level Two	Respond	*React, respond, clarify, reference*
Level Three	Value	*Judge relevance, debate, confront, justify*
Level Four	Organize	*Personalize, defend, prioritize, contrast*
Level Five	Internalize	*Act, display, practice, adopt a behavior*

Fig. 2.7 Bloom's taxonomy of affective domain learning objectives

occurs. In the broadest terms, planetarium learning experiences can occur in formal, school-based learning environments or in informal, out-of-school learning environments. In abbreviated form without any regard to value or hierarchy, planetarium education researchers generally use a short-hand description of learning environments as either being formal or informal (Falk and Dierking 2012).

Learning Environments

Both formal and informal learning environments are deeply critical to the education landscape at large (Feder et al. 2009). Besides the physical location being different between formal and informal learning environments, they also differ to greater and lesser extents in physical size, setting, populations, targeted learning objectives, and ways of assessing their effectiveness. Conducting planetarium education research in these domains can appear vastly different in the methods used for collecting data and how one interprets collected data.

A formal learning environment is typically set within the classroom or at an academic institution, where students are evaluated on a regular basis on their learning progressions by receiving feedback through grades or tests. Traditionally, this type of environment is controlled and structured with instructors "feeding" information to the students though instructional strategies such as lectures, group/individual activities, laboratory demonstrations, etc. During each class, students are expected to participate in scheduled lessons laid out in the syllabus. Additionally, formal environments are set over long, repetitive periods of learning designed to develop a student's understanding through a pre-designed learning progression embedded in the curriculum.

Informal learning environments, in contrast, have a wider range of settings in which they occur, such as museums, after-school programs, nature centers, or social settings (Falk et al. 2007). Feedback received in these types of environments is typically either immediate or does not occur at all. Usually, there is no opportunity for the individual to practice retrieving the information unless they choose to in another setting. Conventionally, these types of environments are unstructured and allow the learner to explore along a path of their choice. Other than being semi-guided by the constraining floor plan of an exhibit hall at a museum, it in no way ensures how much time a learner will spend at each exhibit nor how deeply they will engage with the content.

Population

Let's first consider the population differences. An informal learning environment consists of a wide range of age groups, from pre-kindergarten to those well into retirement. This type of audience creates a challenge for those responsible for engaging such a diverse collection of learners. Each age range has a different set of generational and life-stage interests. For example, a young group of learners could require a more technologically advanced, hands-on type of activity, whereas those in retirement may be more comfortable and able to read a long script of information in silence. In addition, informal environments are charged with the challenge of potentially extremely diverse cultural populations. As important as it is to use a culturally relevant mode of instructional strategies in the formal classroom, it is exceedingly more difficult to define those margins of characteristics in an informal environment where the same features can change daily.

Unlike informal learning environments, a formal setting is expected to have a narrow range within the students' age. The advantage of formal settings is that they can create a curriculum that is specifically designed to target their audience's cognitive abilities. By understanding what is most relevant to students at their age-level, instructors are able to design activities and use instructional strategies that are optimal for engaging their learners. Generally, formal learning occurs between the ages of 5 and 18 years and, for those who choose to continue in higher education, up to approximately 30 years of age. Formal education forces the learner to engage with particular concepts they may not have otherwise. Teachers in formal environments can guide students to the correct answer and provide critical feedback when the learner's current model about a concept is incorrect.

Objectives and Assessment

Formal and informal learning environments also differ in terms of the targeted learning objectives and the evidence they collect for assessment of outcomes. The National Research Council study committee on informal education describes a variety of outcomes that should be central to informal environments, such as pro-moting lifelong, life-wide, and life-deep learning (Feder et al. 2009). Specifically, this study committee created a list of six strands of outcomes from which informal outcomes should operate their framework. The six strands include developing an interest in science, understanding science knowledge, engaging in scientific rea-soning, reflecting on science, engaging in scientific practices, and identifying with the scientific enterprise. In order for researchers to assess whether or not these outcomes are being achieved, the scope of assessment tools can be limited due to the unstructured nature of informal settings. Generally, self-reporting assessments tend to be the most common, if not one of the only ways that researches are able to measure these objectives. Modes of self-reports include, but are not limited to, questionnaires, structured interviews, surveys, focus groups, and talk-aloud protocols.

Informal Setting

In particular, one of the most difficult strands of outcomes to measure in informal environments is the level of understanding scientific knowledge. The reason for this difficulty is due in part to the individualistic nature of this strand and the fact that learners come to the setting with prior knowledge, understandings, and experiences. Evaluating these gains requires thorough pre- and post-tests or interviews, which can be challenging to obtain as well as time consuming.

Due to the challenges that informal learning researchers face, affective variables, such as motivation and interest, are often the primary focus of informal learning research. Researchers such as George Reed, from the planetarium education research community, have found overwhelming evidence in their respective field that the use of the planetarium versus the use of classroom to learn astronomy holds no inherently superior advantages. Influenced heavily by this finding, Reed (1973), and Ridky (1974) advanced the notion that perhaps the greatest value that the planetarium holds is within the affective domain.

However it should be born in mind that these studies from over 40 years ago used a simplified 2D model of the universe projected on the dome. Detailed studies of the advantages of a well-planned classroom activity in a digital planetarium, with its greater ability to use spatial visualization and reinforce astronomical concepts through carefully use of visuals, is an area ripe for future research. One of the earliest attempts to study this employed a sample of students that had the same teacher over many years during which the switch from an optical to a digital planetarium occurred.

Formal Setting

Formal learning environments are generally aimed at maximizing long-term content and cognitive gains for students, as well as developing long-term characteristics that encourage a growth mindset in learners. In order to evaluate these types of outcomes in formal environments, researchers can easily employ nearly any assessment tool including, but not limited to, pre-post-tests, interviews, course evaluations, observations, standardized tests, homework assignments, and case studies. This is usually much easier for the researcher, in that it is a controlled environment that can be altered to incorporate experimental teaching strategies, new curriculum, or activities. Also, the researchers can use two very distinct samples and track them for a longitudinal study. With an informal environment, it can be much more difficult to track participants over a long period of time due to changing phone numbers, emails, addresses, and other contact information. Additionally, planetarium education researchers could run into issues, such as sample bias, since participants in informal environments must partake in the research study voluntarily and be lead by their own motivation. In a formal environment, a sample can be very specific.

Planetariums as a Formal Environment

Formal and informal learning environments, and the planetarium education research that occurs within them, do not always exist as being mutually exclusive, but instead sit along a continuum (Fig. 2.8) (Plummer et al. 2015). Planetarium education research studies often describe some blend of the two and how they interact with one another. For example, 3rd grade students learning about the lunar phases may take a field trip to the planetarium to develop a better understanding. Learning astronomical phenomena can take up a lot of working memory for a novice, and has been shown in research to be superior to the 2-D flat screen computer planetarium programs often used in the classroom or laboratory. Students are technically considered to be learning in an informal environment, but within a formal context. For the *international Studies of Astronomy Reasoning* (*i*STAR) database, Tatge et al. (2016) and colleagues made specific judgments when categorizing studies as formal or informal. Using this particular example, this type of study would be classified as "formal" considering its context. However, say a study was being conducted on an astronomy class for adults held at a museum, which was taught within the planetarium. Technically, a museum is defined as an informal learning institution but a class could be considered more formal. Undoubtedly, the academic lines of distinction between formal education and informal education are unendingly blurred because planetariums in informal settings are often responsible for providing formal-like astronomy instruction to school students.

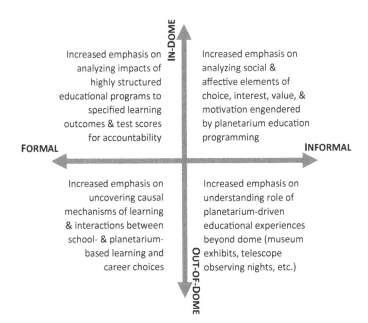

Fig. 2.8 Quadrants of planetarium education research (adapted from Plummer (2015) and colleagues)

Quantitative Methods

Regardless of whether a planetarium learning experience is specifically described as being a formal school-like learning experience or not, the rubber-meets-the-road task for the planetarium education researcher is most often the task of finding some rigorous and dependable way to scholarly describe, quantify, interpret, and explain causal mechanisms about cognitive and affective changes occurring in learners as a direct result of the planetarium learning experience. Planetarium education researchers most often use the time-tested and community accepted methods of empirical research.

Empirical research is generally defined as the scientific way of gaining knowledge through purposeful and systematic collection of evidence via objective observation, experience, and experiment. This type of research serves as the foundation for modern philosophy of science. In the present context, empiricism is a philosophical perspective on understanding the world that is in stark contrast to the differing perspectives of a priori reasoning, intuition, or revelation. Following that line of thinking, the basic empirical research tool kit of planetarium education researchers includes quantitative research methods, qualitative research methods, and mixed-research methods (Slater et al. 2015a).

Unquestionably, the most commonly used research methods in planetarium education research are those from the quantitative research realm. A comprehensive survey by Slater et al. (2016) and colleagues found that 65% of planetarium education research articles and dissertations relied solely on quantitative research methods, another 18% used a mixture of quantitative and other research methods, bringing the total to more than 80% of studies. This is likely a direct result of most planetarium education researchers having at least some formal academic background in the quantitative world of physical sciences like astronomy, which often unconsciously leans heavily on the Lord Kelvin's philosophical axioms that, "If you cannot measure it, you cannot improve it," and "when you cannot express it in numbers, your knowledge is of a meager and unsatisfactory kind".

In an attempt to quantify learning in the cognitive domain and attitudes in the affective domain, planetarium education researchers often use the broad notion of surveys to generate a numerical measure that describes a learner. There are numerous surveys, tests, conceptual diagnostics, attitude inventories, observation protocols, and field-note tabulation schemes available to planetarium education researchers; however, few are widely agreed upon within the planetarium education research community to make study-to-study comparisons easy to accomplish. As a result of too few agreed upon instruments, many planetarium education researchers decide to create their own surveys. Although it might seem easy to create a quick multiple-choice quiz for attendees exiting the planetarium, the accurate construction of a reliable and valid survey is an unexpectedly time-consuming, highly complex, and rigorous task that most planetarium education researchers are academically unprepared to undertake. As a result, the greatest weakness of most planetarium education research studies is the quality, reliability, and validity of the pre-test and post-test survey instruments used. Novice researchers are encouraged to partner with experts for testing and measurement when it becomes necessary to create a new survey instrument.

The *t*-Test

By and large, planetarium education researchers mostly use traditional measures of central tendency to describe their collected data: mean, median, and mode. This allows the mean scores between two measurements—pre-test to post-test gains or post-tests between two groups—to be compared. Across all studies Slater et al. (2016) and colleagues surveyed, the most common statistical tool used in planetarium education research is the *t*-test. The *t*-test is a numerical recipe to determine the likelihood that the difference in test scores between two groups is likely to have occurred by change. For the *t*-test to work, statistical significance is determined by the size of the differences between the groups' mean scores, the sample size of each group, and the standard deviations of the group's scores. It is important to note that statistical *t*-tests comparing mean scores only work consistently when the scores of both groups have similar variance and have a range of scores that are normally distributed.

Qualitative Methods

Qualitative research methods, in contrast, are fundamentally different than quantitative research methods in both their form and underlying philosophical position (Corbin and Strauss 2014; Plummer et al. 2015; Slater et al. 2015b). In recent years, the research method that was once known widely as qualitative research is now more often described as *interpretive inquiry*. Interpretive inquiry is, at its core, a systematic research method to understand how study-participants make sense of educational events and transformative experiences. Planetarium education researchers using this approach analyze interviews and focus group transcripts, make observations of participants' behaviors, analyzing patterns of discourse between learners, and interpret the artifacts of learning, such as student projects, student writings, and student drawings. Unfortunately, novice planetarium education researchers sometimes initially shy away from interpretive research methods because they naively think such approaches are more subjective than quantitative approaches. Experienced researchers understand that there is no truth to the notion that quantitative approaches are more objective and qualitative and interpretive approaches are more subjective. In fact, many Ph.D. dissertations in astronomy education research over the last decade have been mostly qualitative and interpretive in nature (Bailey and Lombardi 2015; Buxner 2015; Slater 2008, 2010).

According to Creswell (2012), the five most common genres of qualitative and interpretive research methods can be categorized as: biography, ethnography, phenomenology, case study, and grounded theory. In brief, biographical research is a method of uncovering the life story and broad influence of an individual. Ethnographic research, in contrast, is focused on conducting observations and interviews to describe and interpret the collective cultural characteristics and experiences of a group of individuals. Phenomenological research is a research method used to explain how individuals or groups understand, experience, and make sense of a shared experience, such as a one-time viewing of a solar eclipse. Case study research methods are used to develop an in-depth interpretation of a historical situation, using interviews or systematic analysis of archived documents. Finally, grounded theory research methods are used when precious little existing information is already known about an individual or group of individuals' learning experiences. The emphasis of grounded theory research methods is to use field data to generate a theoretical explanation that can eventually be tested through experimentation.

These quantitative and qualitative/interpretive research methods are separate and apart from a third research method known as mixed-methods research. Mixed-methods research does not only mean that the planetarium education researcher does some quantitative work and some qualitative and interpretive work. Instead, mixed-methods research is a strategy in which evidence is pursued from multiple angles in order to triangulate conclusions and develop a more complete and accurate picture of the educational learning experience. When considering mixed-methods research, it is worth emphasizing that one common point of misunderstanding among novice planetarium education researchers is when written responses or recorded interview transcripts are collected using the methods of qualitative research, but then words and phrases are counted and numerically tallied to determine the frequency, range, and domain of responses—this strategy has taken initially qualitative research data and converted it into quantitative research,

Table 2.1 Abridged list of journals publishing planetarium education research	Abridged list of journals publishing planetarium education research
	International Journal of Science Education
	Communicating Astronomy to the Public (*IAU*)
	Journal of Astronomy & Earth Sciences Education
	Journal of Geoscience Education
	Latin American Journal of Astronomy Education
	Journal of Research in Science Teaching
	Physical Review—Physics Education Research
	The Planetarian (*International Planetarium Society*)
	School Science and Mathematics
	Astronomy Education Review (no longer being published)
	Journal and Review of Astronomy Education and Outreach (no longer being published)

thereby no longer being qualitative or interpretive research. To be clear, this scenario is not mixed-methods, it would be quantitative methods because the data in this case was converted into quantitative evidence.

Forty years ago, scholarly journals were disinclined to publish anything but research articles solidly based on statistically rigorous quantitative methods. In today's modern era of planetarium education research, all respected journals will publish well-written, carefully done, and timely studies using quantitative, qualitative/interpretive, or mixed-methods research studies. Although there is a misperception among novice planetarium education researchers that only large sample size, multiple-group comparison studies are important and worth publishing, the truth is that that a well-executed case study of just a single person can have as much influence on the field as any other study. A non-exhaustive list of respected scholarly journals that publish planetarium education research is included in Table 2.1.

References

Akey, J. M. (1973). *The behavioral selection of planetarium concepts appropriate for second grade students*. Ed.D. Dissertation, University of Northern Colorado—Greely.

Anderson, L. W., Krathwohl, D. R., & Bloom, B. S. (2001). *A taxonomy for learning, teaching, and assessing: A revision of bloom's taxonomy of educational objectives*. Allyn & Bacon.

Bailey, J. M., & Lombardi, D. (2015). Blazing the trail for astronomy education research. *Journal of Astronomy & Earth Sciences Education, 2*(2), 77–88.

Ball, D. L., & Forzani, F. M. (2007). Wallace foundation distinguished lecture—What makes education research "educational"? *Educational Researcher, 36*(9), 529–540. 2007.

Buxner, S. R. (2015). Exploring how research experiences for teachers changes their understandings of the nature of science and scientific inquiry. *Journal of Astronomy & Earth Sciences Education, 1*(1), 53–68.

Bloom, B. S. (1956). *Taxonomy of educational objectives. Vol. 1: Cognitive Domain*. New York: David McKay, 20–24.

Campbell, D. T., & Stanley, J. C. (1963). *Experimental and quasi-experimental designs for research*. Chicago: Rand McNally.

Corbin, J., & Strauss, A. (2014). *Basics of qualitative research: Techniques and procedures for developing grounded theory*. Sage Publications.

Creswell, J. W. (2012). *Qualitative inquiry and research design: Choosing among five approaches*. Sage.

Chronbach, L. J. (1951). Coefficient alpha and the internal structure of tests. *Psychometrika, 16*(3), 297–334.

Falk, J. H., & Dierking, L. D. (2012). *Museum experience revisited*. Left Coast Press.

Falk, J. H., Storksdieck, M., & Dierking, L. D. (2007). Investigating public science interest and understanding: Evidence for the importance of free-choice learning. *Public Understanding of Science, 16*(4), 455–469.

Feder, M. A., Shouse, A. W., Lewenstein, B., & Bell, P. (Eds.). (2009). *Learning science in informal environments: People, places, and pursuits*. Washington, DC: National Academies Press.

Fisher, M. S. (1997). The effect of humor on learning in a planetarium. *Science Education, 81*(6), 703–713.

Heyer, I., Slater, S. J., & Slater, T. F. (2013). Establishing the empirical relationship between non-science majoring undergraduate learners' spatial thinking skills and their conceptual astronomy knowledge. *Latin American Journal of Astronomy Education, 16*, 45–61. Retrieved from http://web-02.ufscar.br/relea/index.php/relea/article/download/182/248

Krathwohl, D. R., Bloom, B. S., & Masia, B. B. (1964). *Taxonomy of educational objectives: Vol. 2. Affective domain.* New York: David McKay.

Novak, J. D. (1963). A preliminary statement on research in science education. *Journal of Research in Science Teaching, 1*(1), 3–9.

Plummer, J. D., Schmoll, S., Yu, K. C., & Ghent, C. (2015). A guide to conducting educational research in the planetarium. *Planetarian, 44*(2), 8–24, 30.

Reed, G. (1973). The planetarium versus the classroom—An inquiry into earlier implications. *School Science and Mathematics, 73*(7), 553–555.

Ridky, R. W. (1974). *A study of planetarium effectiveness on student achievement, perceptions and retention.* Paper presented at the National Association for Research in Science Teaching, Chicago, IL. Available online at: http://eric.ed.gov/?id=ED091207

Schleigh, S. P., Slater, S. J., Slater, T. F., & Stork, D. J. (2015). The new curriculum standards for astronomy in the United States. *Latin American Journal of Astronomy Education, 20*, 131–151. Retrieved from: http://www.relea.ufscar.br/index.php/relea/article/viewFile/229/313.

Singer, S. R., Nielsen, N. R., & Schweingruber, H. A. (Eds.). (2012). *Discipline-based education research: Understanding and improving learning in undergraduate science and engineering.* Washingon, DC: National Academies Press.

Slater, S. J. (2010). *The educational function of an astronomy REU program as described by participating women.* Ph.D. Dissertation, University of Arizona.

Slater, S. J. (2015). The development and validation of the Test Of Astronomy STandards (TOAST). *Journal of Astronomy & Earth Sciences Education, 1*(1), 1–22.

Slater, S. J., Slater, T. F., Heyer, I., & Bailey, J. M. (2015a). *Conducting astronomy education research: An astronomers' guide, 2nd edition.* Hilo, Hawai'i: Pono Publishing.

Slater, S. J., Slater, T. F., Heyer, I., & Bailey, J. M. (2015b). *Discipline-based education research: A scientists' guide, 2nd Edition.* Hilo, Hawai'i: Pono Publishing.

Slater, T. F. (2008). The first big wave of astronomy education research dissertations and some directions for future research efforts. *Astronomy Education Review, 7*(1), 1–12.

Slater, T. F., & Adams, J. P. (2003). *Learner-centered astronomy teaching: Strategies for ASTRO 101.* New Jersey: Prentice Hall.

Slater, S. J., Tatge, C. B., Bretones, P. S., Slater, T. F., Schleigh, S. P., McKinnon, D., et al. (2016). iSTAR First light: Characterizing astronomy education research dissertations in the iSTAR database. *Journal of Astronomy and Earth Sciences Education, 3*(2), 125–140.

Chapter 3
Learning Research in the Planetarium Prior to 1990

Scholars through the centuries have ardently endeavored to find the best teaching methods to efficiently and effectively transfer vast amounts of accumulated star knowledge to their upcoming generation of sky watchers. Although maps speci-fying the names and locations of stars have been pasted around globes for thou-sands of years, it is the 1923 installation of a large, mechanical star projector in Munich's Deutsches Museum by the Carl Zeiss Company that is most often cited as the first modern projection planetarium for teaching astronomy (Chartrand 1973). In the 1930s, enthusiasm for installing planetariums, or planetaria, spread to the United States. Impressive planetariums were installed in the pre-world War II period, first in Chicago (Adler Museum), to be quickly followed by planetariums in Los Angeles (Griffith Observatory), New York (Hayden at American Museum of Natural History), Philadelphia (Fels at the Franklin Institute), and Pittsburgh (Buhl Planetarium). Just a few years later, Osaka and Tokyo debuted similar planetariums. Planetarium construction increased rapidly in the United States due to federal funding for schools and museums through the 1958 US National Defense Education Act. The US went from one planetarium in 1930, to six in 1940, to about 100 in 1960, increasing to 200 in 1963, 450 by 1967—even before humans had landed on the Moon—and more than 1000 by 1975. Today, somewhere around 3000 per-manent planetarium facilities are available to show the stars to children and adults alike across the world, with perhaps a thousand additional portable planetariums. An annually updated and detailed catalog of planetarium locations can be found through the *International Planetarium Association*'s website.

Since the "first light" emerged from a star projector, planetarium educators have used the systematic rules and logic of experimental scientific investigations to explore, better understand, and ultimately improve, how people learn astronomy in the planetarium. The early planetariums were used for more than just showing enthusiastic guests the names and positions of stars and constellations. Almost immediately, planetariums were used by universities and other teaching entities, including the military, for teaching complex topics of celestial navigation and time keeping at academically high levels. Nicholson (1961) reported that the Hayden

© The Author(s) 2017
T.F. Slater and C.B. Tatge, *Research on Teaching Astronomy in the Planetarium*,
SpringerBriefs in Astronomy, DOI 10.1007/978-3-319-57202-4_3

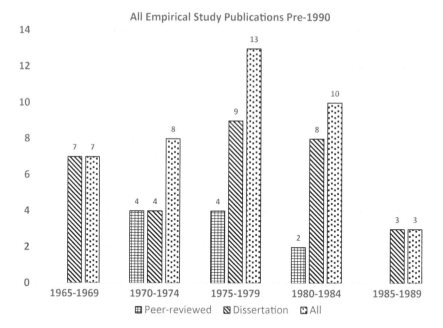

Fig. 3.1 Frequency of peer-reviewed publications and dissertations from 1930 to 1990

Planetarium at New York's American Museum of Natural History offered 19 separate courses in astronomy and navigation.

Although most of these first, large planetariums could seat approximately 100 people at a time, there was an explosion of small planetarium installations made affordable by efficient engineering designs led by Pennsylvania's Armand Spitz (Abbatantuono 1995; Rice and Town 2013). Many of the planetariums were on college campuses and in public school buildings, fueling the vast majority of the scholarly planetarium education research found across a wide spectrum of doctoral dissertations and peer-reviewed journal articles. A timeline of planetarium education research prior to 1990 is shown in Fig. 3.1.

1960's Research

In Washington D.C. (USA), Noble (1964) found that elementary school teachers were largely caught off guard and woefully underprepared by the dawning of the Space Age. This is not surprising since elementary teachers would have had no prior training or education in the nascent Space Age because they were living it, along with their students. As Director of the D.C. School Planetarium, Noble felt responsible for developing an infrastructure to support teachers in becoming better prepared to teach Space Age astronomy and so developed a college-level course for them. Uniquely, and perhaps as the first appearing report in the literature, she taught this class for teachers not in a traditional classroom with desks and chairs, but in the planetarium itself. She judged that Early's (1960) 66-page publication disseminated by the Spitz Planetarium Laboratories entitled, "The Use of the Planetarium in the Teaching of Earth and Space Sciences," was a useful manual for planetarium operators teaching from an Earth-based perspective, but lacked sufficient ability to inform teachers how to best teach from a space-based perspective.

Based on surveys and interviews with five 4th-grade and eight 6th-grade teacher-participants, Noble (1964) found that a planetarium curriculum loosely based on notions of space-navigation helped prepare teachers. The planetarium curriculum involved a simulation of the sky as seen from a space ship in either an equatorial orbit or a polar orbit. The underlying unifying principle, based on work by famed Navy Officer Captain Philip H. Weems who taught navigation to Charles Lindbergh and Fred Noonan, was that it is necessary to imagine the Earth as a lighthouse in the sky against a background of fixed stars and. The "space ship" is then imagined to always be vertically above a point on the Earth's surface. Participants were first taught to locate the Polaris and the orientations of the Big Dipper and Little Dipper from Earth's surface, then from the Moon's surface, from Mars' surface, and finally from a space ship traveling among them.

Nobel's (1964) work was based on looking at how a single group of learners changed as a result of learning in the planetarium, but a more conventional scientific experiment on learning would have been to do a comparison of learning gains between two groups of learners that each participated in different teaching contexts. One of the first two-group experimental comparison studies published in planetarium education was done at Arizona State University. Smith (1966) conducted an experimental comparison of how learning in the planetarium compared to learning in a traditional classroom and measured the results with a teacher-constructed survey. His experimental study compared two groups of 6th grade students using a Campbell and Stanley's (1963) posttest-only, control group study design. He selected elementary level students because 80% of 203 planetariums surveyed by Korey (1963) served as school field trip destinations for elementary students, which is still true today.

Smith's (1966) two comparison groups of students came from 24 classrooms, comprising nearly 700 students. His experimental group of 361 students attended a 40-min planetarium presentation. As a comparison, his second group of 339 students had a traditional 40-min classroom lecture on the same astronomical concepts. Both groups of students then completed the same 25-item multiple choice survey that was iteratively constructed using classical test theory. Smith (1966) found that the posttest average of students visiting the planetarium was 14.21 out of 25 ($SD = 4.34$) in contrast to students who learned astronomy only in a traditional classroom, 16.13 out of 25 ($SD = 3.92$), which was not statistically significant. These results suggest that the planetarium was no better than a classroom presentation, insofar as his test could measure. He also found that students who were taught to identify constellations using projected 35 mm slides—whether displayed in the planetarium or in a classroom—were better able to identify constellations after the lesson than students who were taught constellations using the planetarium sky alone. Although this particular result implies that the projected planetarium sky has significant weaknesses when compared to classroom instruction, the differences in learning may be due to testing students from both groups in a classroom using classroom-like assessments. Smith's (1966) results were consistent with other research done around the same time, including work done by Rosemergy (1967).

Tuttle (1965) advanced the proposition that learning astronomy in the planetarium would not only help students learn astronomy, it would also have the simultaneous educational benefit of increasing students' performance on standardized spatial relations tests. Using two 6th grade classes for a pretest and posttest gain comparison, one class of 26 students learned astronomy in the planetarium while the comparison class of 38 students learned astronomy in the traditional classroom. In both treatments, the instruction stressed areas of visualization and orientation. In a purposeful effort to control confounding variables and to mitigate for possible differences in teachers' teaching ability, the same teacher led the lesson in both the classroom and in the planetarium.

Students who learned in the planetarium increased their pretest scores from an average of 12.2 correct up to an average of 15.2 correct. Students who learned in the classroom did not significantly increase their pretest scores of 12.3 correct to 12.7 correct as compared using a t-test of significance. When looking at subscales, the increase in student scores among those who learned in the planetarium were all related to visualization components of spatial reasoning, whereas the orientation components of spatial reasoning were unchanged regardless of treatment. These results were somewhat similar to work done many years later by Bishop (1980). Tuttle's (1965) study did not directly target how much astronomy students were learning because it was naturally assumed a priori that students would learn astronomy in the planetarium.

Let's take a step back for a moment to consider principles of experimental research design. When carefully scrutinizing the details of these early scholarly reports, most often in the form of doctoral dissertations, the trend in writing about educational experiments of the day when comparing two groups was to describe one group as the *experimental group* and one group as the *control group*. Most often, the study participants that were described as the experimental group was the group that learned in the planetarium, whereas the group that learned in the traditional classroom was usually called the control group. In modern writing, as a stark contrast, planetarium education researchers usually refer to two comparison groups as *treatment-1* and *treatment-2* groups. This is because in a traditional, scientific experimental research study, one of the groups should receive a treatment—in this case an educational intervention—and the other group should receive no educational intervention treatment at all, if it is a control group in the strictest sense. Few of the comparison studies described here use a non-treatment, control group, even though this is how they were most often described.

While one might naturally assume that a planetarium presentation should be innately more effective than a traditional classroom lecture on the same content, Curtin (1967), a planetarium educator at the Detroit Children's Museum, wondered what the actual content of planetarium presentations contained and how planetarium educators' questions would be cognitively classified. About the same time, Ahrendt (1966) and Christin (1967), among others, judged that far too few planetarium educators had sufficient, formal college astronomy training and criticized the larger planetarium education community for having far too weak of a national infrastructure for training would-be planetarium educators. The conventional wisdom for an on-the-job-training planetarium educator was to "visit your neighbors" at other functioning planetariums. Curtin (1967) criticized this approach as creating what he called inbred and "carbon copy" presenters rather than training true educators. This situation caused Curtin to systemically explore to what extent planetarium-visiting school children were actually studying astronomy.

Research in the Great Lakes Region

Given that the US 1958 National Defense Education Act was granting considerable federal funds to school districts in order to construct planetariums (mostly on a matching-funds basis), Curtin (1967) arranged to have planetarium presentations audio tape recorded using the infrastructure of the Great Lakes Planetarium Association. The recordings were then analyzed to determine the astronomy content of school group planetarium programs in an effort to extend nascent work done by Richard Emmons (1950) and Anita Emmons (1963). A content analysis of 38 audio tapes was conducted, as well as a cognitive, understanding-level domain classification of questions posed to students by planetarium educators; these questions were written based on Bloom's Taxonomy of Cognitive Objectives. He studied 18 elementary school program recordings and 20 junior high school programs. He reported that 42.9% of the presentation time at the elementary school level was allocated to current, night sky observations, and that 48.5% of the presentation time for junior high school level programs were devoted to current, night sky observations. In this sense, about half of the presentation time was talking about constellation location and star-planet location.

For his sample of 38 programs, elementary programs averaged 45.3 min and junior high programs averaged 47.4 min. He suggests that the difference is a result of a longer "lobby talk" about rules for behavior in the planetarium to elementary classes before entering the theater, or simply longer introduction times in general. In terms of scientific content, Curtin (1967) found that an average of 9.8 constellations was taught to elementary students whereas 14.7 constellations were taught to junior high students in each program. The most popular constellations and asterisms were the Big Dipper, the Little Dipper, Leo, Taurus, and the constellations connected to the Summer Triangle. A second highly common group of constellations frequently showing up in programs were those connected to a Greek mythical story uniting Andromeda, Perseus, Cassiopeia, Cepheus, and Pegasus.

Table 3.1 Objects taught with high frequency—Adapted from Curtin (1967)

Proper Name	Elementary Level ($n = 18$)	Junior High Level ($n = 20$)	Total Occurrences ($N = 38$)
Polaris	15	15	30
Saturn (*not a star*)	10	13	23
Moon (*not a star*)	16	7	23
Sirius	8	10	18
Castor & Pollux	8	6	14
Betelgeuse	4	8	12
Vega	2	9	11
Altair	0	10	10
Capella	4	5	9
Aldebaran	3	6	9
Spica	3	2	5
Alpha Centauri	0	3	3

A listing of specific objects that were taught with relatively high frequency is shown in Table 3.1. The most commonly mentioned object was Polaris, but perhaps surprisingly, it was not mentioned in all 38 programs. The Moon and planet Saturn was the most common non-stellar object mentioned, which is perhaps unexpected as Saturn isn't often visible in the night sky. The brightest star in the night sky, Sirius, was only mentioned 18 of 38 times. Curtin (1967) also noticed common phrases that were frequently repeated to all age groups and across all facilities. These are listed in Table 3.2.

While the items listed in Tables 3.1 and 3.2 might initially seem reasonable for a planetarium presentation, it turns out that this represents only an extremely small fraction of the breadth of astronomy expected to be taught in schools of the day. Although there is no nationally mandated science curriculum in the United States, one method to establish a baseline is to consider the astronomy and space science topics included in the widely adopted *Earth Science Curriculum Project* (ESCP)

Table 3.2 Frequent planetarium educator statements identified by Curtin (1967)

1. Mercury is the closest planet to the Sun
2. The apparent motion of the stars is due to the daily motion of Earth
3. The atmosphere of Earth helps prevent our seeing stars in the daytime, and causes stars to appear to twinkle at night
4. Revolution of Earth is a cause of seasons; rotation is a cause of day and night
5. Mars appears to be red in color
6. Jupiter is the largest planet and it has twelve [*sic*] moons
7. Saturn has rings
8. Pluto is the farthest planet [*sic*] from the Sun

Note These are not listed in any particular order

guidelines for curriculum learning objectives (Irwin 1970). The ESCP has only the slightest mention of naming stars and constellations as part of its much broader astronomy and space science lists of topics to be taught. As a second issue of concern, Curtin (1967) reports NASA's Gemini space program as being mentioned more than any other ongoing space program, and then only 11 times invariably when describing the constellation of Gemini. This alarmingly suggests that the planetarium programs he surveyed were neither serving crucial school curriculum goals nor enhancing public appreciation and understanding of national scientific and engineering goals the 1958 National Defense Education Act was trying to enact.

The frequency of questions being posed in the planetarium learning environment is frequently considered as an indication of the level of student interactivity and engagement in learning environments, with few to no questions being asked as indicative of a passive-student learning environment. He found that, independent of cognitive level, non-rhetorical questions were posed to students in 14 of 18 elementary programs and in 17 of 20 junior high programs. In terms of measuring the educational rigor of planetarium programs, Curtin (1967) attempted to quantify this by categorizing the frequency of high-level cognitive questions posed to attending students by planetarium educators in his sample. He was able to identify 413 questions across the 38 programs he studied. Almost all questions posed were of the lowest-cognitive memory level, with only 9 higher-level cognitive questions posed to students. This calls into question the didactic teaching skills of the planetarium educators lecturing in the planetarium.

Taken together, Curtin's (1967) research was alarming in that it revealed a mismatch between planetarium program content and the ESCP guidelines as well as a mismatch between the cognitive level of planetarium presentations and those addressed by skilled teachers. However, this work was done by considering the planetarium program in isolation of other learning experiences that might occur before or after the planetarium program in the larger landscape of students' overall portfolio of learning experiences. As a result, several years later, Reed (1972) formally outlined for the community the knowledge and skills the next generation of planetarium educators would need in order to be successful.

Student Achievement Research

At about the same time Curtin (1967) was working, Wright (1968) was working at the University of Nebraska trying to compare 8th grade student achievement on astronomy topics. Wright studied differences among groups of students who had various levels of pre-planetarium visit or post-planetarium visit instructional learning experiences. One might assume that student learning would be enhanced either by more time-on-task in studying astronomy or if the planetarium trip was accompanied by supportive classroom learning experiences that were aligned with planetarium educators' teaching goals. The educational value of pre- and post- field trip learning activities was well established at that time in a number of different academic domains (Chonholm 1955; Evans 1958). Yet, there was little to no highly regarded empirical research to support the widespread notion of the planetarium as being highly effective in teaching astronomy and space science; nonetheless, a 1963 issue of *The Science Teacher* magazine devoted an entire issue to the importance of planetariums for reaching national science education goals.

Wright (1968) conducted a complex study of student performance on an expert-review validated 80-item true-false posttest astronomy knowledge survey among 1600 8th grade students who were divided into four treatment groups and scored as follows: (Group 1) Completed an astronomy unit and had not yet attended the planetarium program, $n_1 = 737$, $x_1 = 24.7\%$ correct; (Group 2) Completed an astronomy unit and attended a planetarium program, $n_2 = 450$, $x_2 = 30.9\%$ correct; (Group 3) Completed an astronomy unit, participated in special preparation conducted by the classroom teacher, attended a planetarium program, and completed a follow-up activity, $n_3 = 218$, $x_3 = 28.5\%$ correct; and (Group 4) Completed an astronomy unit, participated in special pre-visit preparation activity conducted by the planetarium lecturer, attended a planetarium program, and then completed a follow-up exercise, $n_4 = 196$, $x_4 = 30.1\%$ correct. These results are summarized in Table 3.3.

Wright (1968) conducted a standard practice, one-way analysis of variance (ANOVA). The analysis suggests that the most significant difference is that the students attending the planetarium lesson scored higher on the posttest in a statistical sense, although the effect size in this study was quite small. The analysis also questions the required energy and expense to build a planetarium and conduct travel to the facility. At the same time, the various preparation and follow-up activities did not appear to dramatically improve students' achievement insofar as the instrument could measure.

Table 3.3 Summary of Wright's (1968) results

Group	Instructional treatment	Posttest Percent Correct (%)
$n_1 = 737$	Completed classroom astronomy lesson but did NOT visit the planetarium	24.7[**]
$n_2 = 450$	Completed an astronomy unit and attended a planetarium program	30.9
$n_3 = 218$	Completed an astronomy unit, participated in special preparation conducted by the classroom teacher, attended a planetarium program, and completed a follow-up activity	28.5
$n_4 = 196$	Completed an astronomy unit, participated in special pre-visit preparation activity conducted by the planetarium lecturer, attended a planetarium program, and then completed a follow-up exercise	30.1

Note **Indicates statistically significant difference at the $\alpha = 0.01$ level

Reed's Research—1970s

Because of unresolved concerns about the lack of valued scientific content (Smith 1966), questions about the skills of the planetarium educators conducting the programs (Curtin 1967), and the seeming irrelevance of pre- and post-instructional visits on student achievement (Wright 1968), Reed (1972) designed several studies to compare high-quality classroom instruction with high-quality planetarium education. He employed carefully selected scientific concepts and the best practices in instruction. Working at West Chester State College near Spitz Planetarium Laboratories, Reed measured the effectiveness of the planetarium as a teaching facility as compared to a classroom instructional setting relative to clearly specified behavioral objectives. Although not the first planetarium education researcher to do a two-group comparison study between the planetarium and the classroom, his classroom-based instructional strategy was unique because he made significant use of student-manipulated celestial spheres during instruction. His approach of making students more active learners in the process increased the quality of the non-planetarium instruction significantly.

Reed (1970, 1972, 1973, 1976) specifically measured students' understanding of diurnal and yearly motions of stars, motions of the superior planets and the sun, and precession using an expert validated test that had undergone several iterative rounds of pilot testing. In those days, planetarium education researchers most commonly used the posttest-only, control group design because they were unsure of the extent to which knowledge was inadvertently gained from a pretest. Note that this would be a confounding variable in the study. In addition to a post-instruction posttest, he also administered a post-posttest 8 weeks after instruction to determine the durability and retention of students' understanding as a result of the instructional treatments. Reed's (1973) data on 159 college students suggested that there was no statistically significant difference in student scores in the two treatments.

As a planetarium educator of international reputation, Reed himself was highly influential across the planetarium education field for four decades. Based on his work, he interpreted his results as having two longstanding implications. The first is that for teaching, the planetarium theater itself is best used not as a passive theater but when used as a teaching classroom, and using all the best teaching practices and pedagogical techniques available for teaching children. The second longstanding notion was that the true value of the planetarium might lie in the affective domain, rather than the cognitive domain, for enhancing students' interests and attitudes toward astronomy, and science in general.

Asking the Right Questions

Also working in the early 1970s at the University of Northern Colorado, Akey (1973) was interested in conducting a systematic evaluation of astronomy concepts taught in the planetarium to determine the most appropriate for young children.

With a study group of 150 2nd grade students in the Midwestern US, he investigated the extent to which 56 concepts (presented across three separate but sequential planetarium programs) could be retained after a two-week time lapse using a single-group, pretest-posttest study design and evaluated by a Chi-square test of statistical significance. To Akey's surprise, student-participants already held appropriate understanding of 24 of the 56 identified concepts at the beginning of the study on the pretest. After the planetarium programs, a substantive number of students understood nearly all of the concepts. Based on his results, he recommends that the notion of "pointer stars to indicate the position of Polaris" and "the length of a year for superior planets" as being too cognitively complex to durably teach 2nd graders. He thus recommended that they be excluded from future 2nd grade teaching objectives.

Today, planetarium educators find that Akey's (1973) study lacks face validity. For one, the survey questions for students seem to require high levels of reading ability to answer, especially when considering that they were taught through a lecture in a planetarium. Questions included, for example: "Circle the name for the period of revolution of the earth around the sun, from the following: day, month, year" and "Circle the names of a planet that has a shorter year than Earth, from the following: Mars, Mercury, Venus." Many educators would consider test items using this particular reading-level to be too high for contemporary 2nd grade students, bringing into question the validity of Akey's test. As a second criticism, most educators today would probably consider 56 concepts to be an unimaginably large number of concepts to try to teach to 2nd graders. Finally, the notion that so many students would be able to answer the majority of these questions correctly seems implausible in concert with other studies conducted during the same era. His results were similar to those of Sunal (1973) from the University of Michigan who also studied 2nd grade students and also found, using a seemingly complex 30-item, multiple-choice test, that 986 students knew a surprisingly large amount even before attending planetarium. However, there was no statistically significant posttest difference measured when comparing learning in the planetarium to learning in the classroom.

Need for Standard Instruments

The broader community of planetarium educators was naturally dismayed that education research consistently failed to conclusively demonstrate the power of the planetarium learning environment to dramatically impact student achievement. One possible reason forwarded for the lack of positive learning results was that each researcher had to make his or her own knowledge survey instrument. The creation of an instrument that is widely valid and highly reliable across institutional contexts is generally regarded as a highly difficult endeavor (Slater et al. 2015). In short, there existed no widely agreed upon way to measure student achievement, nor any way to easily compare student learning in different facilities. In order to have standardized instruments so that researchers could better compare results among their studies, several efforts to create community-wide knowledge tests were undertaken.

At Michigan State University, Bondurant (1975) created a constellation identification test designed to be used with upper elementary students. At Georgia State University, Hayward (1976) designed a survey for quantitatively assessing students' conceptual understanding of annual motion. These were both similar to earlier efforts by Guilbert (1972) who validated a college-level standardized test specifically for the planetarium and by Battaglini (1971) who advocated that the planetarium itself could serve as a more natural testing environment. In the end, there was no consensus within the planetarium education community and, to date, there is no ubiquitously accepted planetarium education knowledge survey instrument.

The Planetarium Environment

About the same time, Ridky (1973, 1974, 1975) was conducting comparative pretest-posttest learning experiments with both junior high school students and college students visiting the planetarium. He was interested in determining how to most effectively structure the overall planetarium learning experience. In those days, as it is today, a field-trip to the planetarium was often a dramatic experience for students who normally spent their days seated in school desks aligned in straight rows facing a teacher and a chalkboard. A trip to the planetarium was awesomely different. A common practice among planetarium educators that is still true today is to usher student groups into the planetarium. The students are directed to sit in reclined seats that look nothing like school desks, under a hemispherical shaped ceiling where the lights are low, perhaps red in color, and space-age ethereal music plays in the background. Such a classroom is about as different of a learning environment for students as one could imagine.

Ridkey (1975) wondered if some of the unusual aspects of the physical learning environment were somehow getting in the way of students learning about the heavens. In other words, he hypothesized that the uniqueness of the planetarium classroom was sufficiently distracting students' cognitive abilities and was impairing student learning. He designed a two-group comparison study of 52 junior high students and used a 20-item multiple choice posttest covering concepts of celestial motions to measure students' learning in the planetarium under two treatment conditions. In one treatment, students were brought into the darkened planetarium in the usual way to experience a planetarium presentation. In a second, comparison treatment, students were brought into a well-lit planetarium with the lights fully turned on. Before the lights were dimmed, the planetarium educator took a few minutes to describe why the seats were reclined, why they would be turning off the lights, why the ceiling was hemispherical, and, in general terms, how the odd-looking giant star ball projector in the center projected the appearance of rising and setting stars on the ceiling. After a brief explanation, the lights were dimmed and the planetarium presentation continued in the typical way.

A comparison of the two groups' posttest scores showed that the students who were told how the planetarium works scored significantly higher than students who were not shown how the planetarium works. He repeated this experiment with college-level students and found the same result. Ridky (1974) called this phenomenon the "mystique effect" and it was one of the most important planetarium education research findings of the decade; however, his work was not well publicized throughout the planetarium education community. Even today, the community tries to enhance the mysteriousness of the planetarium theater by shielding how it works from audiences, despite the apparent negative consequences on learning due to the cognitive distractions caused by the unusual environment.

Several years later, Fletcher (1977) designed an experimental comparison study at the University of Virginia to determine the educational impact on student achievement in a traditional planetarium lecture and in a planetarium lesson with student participation. His study was motivated by a rapidly growing predilection across the science educational research community that active student engagement was required for enhanced learning of scientific concepts (viz., Ausubel et al. 1968; Doyle 1977; Gagne' 1973; Karplus 1980; among others). In other words, along the lines advocated by Reed (1973) and others, there was some growing cultural pressure within the planetarium education community to use interactive teaching approaches rather than traditional information-download lectures in the planetarium (Schatz and Friedman 1976). However large the theoretical literature base supported using this new teaching strategy, there was little to no experimental research in the planetarium to support this position.

Interactive Versus Lecture-Based Comparison

Fletcher (1977), who was president of the Southeastern Planetarium Association, used a multiple-group, multiple-measures posttest study design by administering a 20-item, multiple-choice posttest knowledge survey. This survey was meant to study the impact of two different style planetarium programs: interactive participation-based programs and traditional lecture-based programs. Whereas the traditional treatment was a lecture, the experimental condition asked students to complete fill-in-the-blank worksheets during the planetarium program. The planetarium educator would pause each time students needed to complete an entry on their worksheet. This strategy was consistent with the contemporary notion of active learning of the day, and is in stark contrast to the more common learning experience where students sit passively and listen to the planetarium educator point out stars, constellations, and their predictable motions.

Nearly 1000 student-participants, about half of which experienced each treatment, took the posttest immediately after visiting the planetarium and then again four weeks later, using machined gradable *Scantron* bubble-sheets. Uniquely, for a planetarium educator at the time, he divided his multiple-choice survey items into two levels: Sixty percent of the 20 items were memory-level recall questions and 40% of the 20 items were designed to be higher-cognitive level application questions. His analysis showed that there was no difference in achievement scores for either type of question among any of the treatment groups insofar as his instruments could measure. He did observe relatively high variances between groups, which he speculated were due to differences in planetarium educator skills. Today, planetarium educators would likely attribute this to differences in the student samples, rather than variations in planetarium educator skills. Either way, Fletcher (1980) was not able to definitively establish that his treatment groups were equivalent samples at the beginning of the study and the end result is the same—there was no measureable difference in the two treatment methods.

The disappointing null results in terms of achievement gains seen by Reed (1970), Ridky (1974), Fletcher (1977, 1980) and others were not immediately confirmed by all planetarium education researchers engaged in similar comparison studies. Ortell (1977) compared the homework, exam scores, and end-of-term final grades of his 241 Cerritos College and El Camino College students and disaggregated the data according to who attended class in the planetarium. Of the 135 planetarium attending college students, 28% received grades of A, 40% grades of B, 15% grades of C, and 15% grades less than C. Of the 106 non-planetarium attending students, 11.3% had grades of A, 30% grades of B, 25% grades of C, and 34% grades less than C. He further disaggregated the data by gender, age group, and overall grade point average (GPA). For all subgroups, he reports that college students who attended class in the planetarium earned higher grades than students who did not. These conclusions have been widely criticized because he had few controls on confounding variables, and was further unable to mitigate for possible differences in teaching methods in the classroom to teaching methods in the planetarium (viz., Edoff 1982).

Nonplused by the seeming contradiction between using good teaching practices in the planetarium and the lack of demonstrably higher achievement, Mallon and Bruce (1982) devised a large scale, multi-facility study to compare learning in lecture-centered, night sky, star-tour shows for more participatory-oriented planetarium learning experiences. He was motivated to focus on small, 15–75 person capacity planetariums. 96% of the United States' 1100 facilities were small facilities that focused predominately on school-aged attendees more than the larger museum-based facilities.

Stronger Evidence in 1980s

Described in detail by Mallon (1980), who was a doctoral student at Temple, five locations distributed across the United States participated in this study, targeting 556 8–10 year old students: 324 from Methacton, PA; 76 from Richardson, TX; 52 from Berkeley, CA; 52 from St. Paul, MN; and 52 from Reno, NV. Two common

Table 3.4 Behavioral objectives targeted by Mallon and Bruce (1982)	1. Define the term 'Constellation' as a group of stars connected together to form a shape in the sky;
	2. State that constellations can be used as 'Skymarks' or signposts to locate other objects in the sky (e.g., nebulae, planets);
	3. Recognize a given set of constellations as represented on a 'Star Map';
	4. Demonstrate the use of a 'Star Map' in locating an object or constellation in the sky; *and*
	5. Relate various pieces of factual information about the sky (e.g. if a star is seen next to one star on one night, it will be next to that star on other nights)

40-min program scripts that used both the star projector and a 35 mm slide projector were developed that focused on five behavioral learning objectives; this is listed in Table 3.4. A 22-item, multiple-choice, paper- and -pencil test was distributed and read aloud to student-participants. Simultaneously, an overhead projection of each test item was shown to students to support the widest possible degree of understanding of each item. Correlated *t*-tests were used to compare pretest and posttest scores and two-way factorial analyses of variance compared the groups' posttest scores. On average, students who attended the traditional lecture scored about 60% correct, whereas students who attended the participatory oriented program, characterized by a vaguely described activity-based format and extensive verbal interaction, scored about 75% correct on the posttest (pretest scores were not reported) across all five study sites. In contrast to earlier studies, Mallon and Bruce (1982) concluded that their data provided clear evidence that a small planetarium facility could be most effective at teaching constellations when used in an active-learning classroom mode.

Bishop's Research

About the same time, Bishop (1980) designed a study to determine if the nascent three-phase *Learning Cycles* approach to active learning prescribed by Karplus (1980) might result in more robust evidence that participatory planetarium programs were more effective at teaching students than traditional planetarium lectures. Karplus (1980) suggested that students would learn more science if teachers purposefully led them through a three-phase sequence of learning experiences. Described eloquently by Rischbieter et al. (1993), the first phase, which Karplus named *exploration*, was to allow students to engage in an unfamiliar scientific phenomenon in a discovery-mode. In this mode, the teacher's specific learning goals were not revealed to the students who were encouraged to use informal language to catalog and describe their observations. The second phase, called *concept introduction*, was a teacher-led instructional experience where teachers helped students identify salient features of a phenomena and use more scientifically accurate vocabulary. The third phase, known as *concept application*, provided students with novel situations to analyze using the language, understanding, and skills they developed in the previous phase. This three-phase learning cycle eventually influenced the widely adopted, modern, five-phase learning sequence widely known today as Bybee's (2014) 5E model.

In summary, Bishop (1980) questioned whether students learned more if they could have repeated experiences with observations, manipulate models, and make scientific sketches, much as students might in Reed's (1972) vision of planetarium as a circular classroom.

Bishop (1980) designed an extensive two-group, posttest only, control group experimental design. The typical sampling approach of the day was to divide students into treatment groups based on who their teacher was, but Bishop used a randomized sampling approach. 98 8th grade students from Northeastern Ohio, who were all taught by the same teacher, were randomly assigned to either the experimental, learning cycle-style planetarium learning experience or to a traditional planetarium lecture treatment. Whereas many of the previous research survey instruments could be classified as lower-cognitive level memory and recall objectives, the learning targets Bishop specifically chose to study were those higher-cognitive level ones that she believed required substantive spatial reasoning ability: motions of the celestial sphere, Earth's rotation, seasons, lunar phases, and predictable planetary motions. Students in both the experimental and traditional treatment groups attended the planetarium eight times and immediate and delayed astronomy posttests were given to the student participants. In addition, students completed the *Differential Aptitude—DAT Space Relations Test* at the beginning and end of the study to assess their spatial ability to look for correlations (Bennett et al. 1974).

Because of the extended duration of 8 instructional sequences, in concert with detailed task analyses for traditional and experimental teaching conditions, and the post-posttest astronomy knowledge surveys along with pre-post spatial reasoning assessments, Bishop (1980) was able to explore 29 different hypotheses about student learning. Rather than list them all here, her hypotheses can be roughly organized into several broad categories:

 (i) hypotheses related to comparing traditional and active-learning oriented planetarium program effectiveness for overall learning;
 (ii) hypotheses related to the comparison of traditional and active-learning oriented planetarium program effectiveness for learning specific conceptual ideas;
(iii) hypotheses related to impact of students' gender in the various learning conditions;
(iv) hypotheses related to differences in learning among high and low-achieving students.

After considerable analysis, she was able to find compelling data in support of participatory planetarium approaches where students manipulated models such as celestial spheres, made sketches describing their observations, and were involved in Socratic-style question and answer dialogues with the teacher. In other words, the concept of the planetarium being an interactive and multi-modal participatory was shown to be educationally superior to a conventional planetarium lecture where students were passive observers or even those planetarium presentations where students were asked questions about what they were seeing. These results were consistent with the interactive notions educational theoreticians of the day were advocating. It seems that earlier participatory planetarium learning experiences were simply not sufficiently participatory to see the achievement that Bishop (1980) was able to produce.

Active Participatory Learning

As specific prescriptions about what exactly "active participatory learning" looks like in science classrooms became more widespread, some innovative planetarium educators began the process of adopting the best practices in science instruction into the planetarium learning environment. This is not to say that all, or even most, planetarium educators were using best practices in science teaching in their planetariums. In fact, most public school science teachers were not using these emerging, educational theory-driven and research-confirmed, best practices; instead, they were relying on long-standing traditional instructional models of lecture-based, information download teaching approaches. In the same way, the majority of planetarium educators were using long-standing lecture approaches to teach in their planetariums. Despite overwhelming educational research to the contrary, outdated and ineffective teaching approaches are still the most commonly used today, as they were in the 1980s, in both formal classroom and planetarium educational settings.

One of the rapidly adopted approaches to improving learning in formal science classrooms was the use of *advance organizers* (Ausubel 1960, 1978) to help students learn by providing them with a mental map, or mental schematics, to which students could organize new scientific ideas they were learning as they were encountering the concepts (Fig. 3.2). Another teaching strategy gaining popularity was that of *clustering*. Bousfield and Bousfield (1966) found that students had a more accurate recall of long lists of facts when they were taught together as "clusters". In the science classroom, this might manifest itself as a teacher leading students through memorization exercises that emphasize sequences of names, rather than individual entities. An example of clustering names together in a sequence from the planetarium education is to have students memorize the phrases, "follow the arc of the *Big Dipper*'s handle to *Arcturus*, then spike right on down to *Spica*"

Fig. 3.2 Example of an
advance organizer, adapted
from Marzano et al. (2001,
p. 119)

Students in the 6th grade were going to take a field trip to the planetarium.For homework, Mr. Armstrong asked student to look at photocopies of a star atlas showing which stars and which constellations will be visible. He said, "let's become familiar with some of the names and patterns we will be seeing in the planetarium so that you will know what you are looking for when we get there."

and "the winter stars make an alphabetized circle surrounding the constellation of *Orion*: an *Aldebaran, Capella, Castor, Pollux, Procyon*, and *Sirius*."

Giles (1981) at Pennsylvania State University designed a study to see how advance organizers and clustering teaching strategies might play out in a planetarium learning environment. The study sought to compare the effectiveness of advance organizers and clustering, both singly and in combination, upon learning in the planetarium. He designed a multiple-group, pre-test posttest control group study design with 832 non-randomized participants. Essentially, he had three different experimental treatments, and a fourth group he called the control group to serve as a comparison group. These groups were: *Treatment 1*-the control group who attended a traditional lecture presentation in the planetarium; *Treatment 2*-instruction in the same subject matter with clustering; *Treatment 3*-instruction in the same subject matter but with advance organizers provided by the planetarium educator; and *Treatment 4*-instruction in the same subject matter with both clustering and advance organizers.

Giles (1981) used an ANOVA statistical test of significance to evaluate student achievement from the pre-test to posttest achievement gains on a 20-item multiple choice knowledge test. He found that although his student-participants' posttest scores were lower than he would have liked, there were some statistically significant differences among the groups. The control group who experienced a traditional planetarium lecture scored 43% correct, which was significantly lower than all other treatments. The group who was taught in the planetarium using memory clustering techniques scored 50.1% correct, and the group taught in the planetarium using advance organizer strategies scored 50.3% correct. These two treatment groups scored statistically the same. Finally, the group taught in the planetarium using a combination of memory clustering techniques and advance organizer strategies scored 60.0% correct, which was statistically higher than all other groups. His conclusion was that students would greatly benefit from as many theoretically-driven, research-confirmed teaching strategies available to planetarium educators that could be included in planetarium lessons.

Planetarium education researchers were now consistently finding that the planetarium by itself is not a miracle machine for engendering learning. The planetarium instructional interventions needed to adopt the same best practices for effective science teaching that successful classroom teachers were using. To further explore this idea, Edoff (1982) conducted a posttest only, two-group comparison study of 542 5th and 8th grade students to determine if using manipulatives under the planetarium dome would enhance students' achievement. Hands-on teaching manipulatives, such as the celestial sphere used by Reed (1970), were gaining considerable attention across the science education landscape as being helpful in classroom science instruction across the disciplines. In his study, both groups of students learned astronomy in the planetarium, but only one group did so by individually manipulating learning materials. He studied three big ideas in astronomy: (i) celestial sphere and time (where only one group used plastic hemispheres to mark changing positions of sky objects); (ii) seasonal changes (where only one group marked paths of the Sun on plastic hemispheres; and (iii) lunar movement and moon phases (where only one group used a flashlight and small spheres to simulate phases). He found that the proper use of manipulatives resulted in greater learning gains for students who used them.

Progress—the First Thirty Years

Taken together, the first thirty years of planetarium education research evolved considerably. Early studies were focused on attempting to demonstrate that the planetarium could live up to its promise as an education-ready, high-tech solution to the nation's needs to better educate students who would presumably go on to fill the science and engineering career pipeline. When planetarium education researchers by and large failed to demonstrate easy-to-achieve learning gains compared to traditional classroom instruction, researchers turned to better understanding the underlying mechanisms that encouraged learning. The planetarium ultimately became a useful tool in the broader science teaching portfolio of the United States, despite some early failures. This turn for the better occurred when planetarium educators were able to appropriately adopt the educational theory-driven and research-confirmed practices into science teaching, and incorporated these into the planetarium learning environment.

References

Abbatantuono, B. P. (1995). Armand Spitz-Seller of Stars. *The Planetarian, 24*(1), 14–22.

Ahrendt, M. H. (1966). *Planetarium survey.* Paper presented at the Great Lakes Planetarium Association Meeting at the Cincinnati Museum of Natural History, October 14, 1966, Cincinnati, OH.

Akey, J. M. (1973). *The behavioral selection of planetarium concepts appropriate for second grade students.* Ed.D. Dissertation. Greely: University of Northern Colorado.

Ausubel, D. P. (1960). The use of advance organizers in the learning and retention of meaningful verbal material. *Journal of Educational Psychology, 51*(5), 267.

Ausubel, D. P. (1978). In defense of advance organizers: A reply to the critics. *Review of Educational Research, 48*(2), 251–257.

Ausubel, D. P., Novak, J. D., & Hanesian, H. (1968). *Educational psychology: A cognitive view.*

Battaglini, D. W. (1971). *An experimental study of the science curriculum improvement study involving fourth graders' ability to understand concepts of relative position and motion using the planetarium as a testing device.* Ph.D. Dissertation. Michigan State University.

Bennett, G. K., Seashore, H. G., & Wesman, A. G. (1974). *Differential aptitude test* (5th ed.). New York: Psychological Corporation.

Bishop, J. E. (1980). *The development and testing of a participatory planetarium unit employing projective astronomy concepts and utilizing the karplus learning cycle, student model manipulation and student drawing with eighth-grade students.* Doctoral Dissertation. University of Akron.

Bondurant, R. L. (1975). *An assessment of certain skills possessed by fifth-grade students used to successfully identify constellations in a planetarium.* Ph.D. Dissertation. Michigan State University.

Bousfield, A. K., & Bousfield, W. A. (1966). Measurement of clustering and of sequential constancies in repeated free recall. *Psychological Reports, 19*(3), 935–942.

Bybee, R. (2014). The BSCS 5E instructional model: Personal reflections and contemporary implications. *Science and Children, 51*(8), 10–13.

Campbell, D. T., & Stanley, J. C. (1963). *Experimental and Quasi-Experimental Designs for Research on Teaching.* American Educational Research Association. Chicago: Rand-McNally.

Chartrand, M. R. (1973). A fifty year anniversary of a two thousand year dream: The history of the planetarium. *Planetarian, 2*(3), 95–101.

Christian, J. (1967). Getting acquainted with the GLPA. *Great Lakes Planetarium Association Newsletter, 4,* 1.

Cronholm, L. C., & Lansing, E. A. (1955). what are the educational values in planned field trips? *NASSP Bulletin, 39*(210), 88–91.

Curtin, J. T. (1967). *An analysis of planetarium program content and the classification of demonstrators' questions.* Doctoral Dissertation. Wayne State University.

Doyle, W. (1977). Paradigms for research on teacher effectiveness. *Review of Research in Education, 5,* 163–198.

Early, R. N. (1960). *The use of the planetarium in the teaching of earth and space sciences.* Chadds Ford, PA: Spitz Laboratories. ASIN: B001KSTVMO.

Edoff, J. D. (1982). *An experimental study of the effectiveness of manipulative use in planetarium astronomy lessons for fifth and eighth grade students.* Ed.D. Dissertation. Wayne State University.

Emmons, A. J. (1963). *A Study of Planetariums and Planetarium Programming.* Ohio: Senior Honors Project, Kent State University.

Emmons, R. H. (1950). *A report on a school planetarium: Its design, its development as a group project, its utility as an instructional aid and its program in school community relations (Master's thesis).* Kent State University.

Evans, H. C. (1958). *An experiment in the development and use of educational field trips.* Ed.D. Dissertation. University of Tennessee.

Fletcher, J. K. (1977). *An experimental comparison of the effectiveness of a traditional type planetarium program and a Participatory Type Planetarium Program.* Doctoral Dissertation. University of Virginia.

Fletcher, J. K. (1980). Traditional planetarium programming versus participatory planetarium programming. *School Science and Mathematics, 80*(3), 227–232.

Gagné, R. M. (1973). Learning and instructional sequence. *Review of Research in Education, 1,* 3–33.

Giles, T. W. (1981). *A comparison of effectiveness of advance organizers and clustering singly and in combination upon learning in the planetarium.* Doctoral Dissertation. Pennsylvania State University.

Guilbert, E. H. (1972). *A standardized test in collegiate descriptive astronomy on selected concepts which can be demonstrated in the planetarium.* Doctoral Dissertation. University of Southern Mississippi.

Hayward, R. R. (1976). *The developing and field testing of an instrument using the planetarium to evaluate the attainment of the concept of annual motion.* Paper presented to the National Association for Research in Science Teaching (*based on his Doctoral Dissertation from Georgia State University*), April 23, 1976, San Francisco, CA.

Irwin, L. A. (1970). A brief historical account and introduction to the Earth Science Curriculum Project. *The High School Journal, 53*(4), 241–249.

Karplus, R. (1980). Teaching for the development of reasoning. *Research in Science Education, 10* (1), 1–9.

Korey, R. A. (1963). *Contributions of planetariums to elementary education.* Doctoral Dissertation. Fordham University.

Mallon, G. L. (1980). *Student Achievement and Attitudes in Astronomy: An Experimental Study of the Effectiveness of a Traditional "Star Show" Planetarium Program and a "Participatory Oriented Planetarium" Program.* Ed.D. Dissertation. Temple University.

Mallon, G. L., & Bruce, M. H. (1982). Student achievement and attitudes in astronomy: An experimental comparison of two planetarium programs. *Journal of Research in Science Teaching, 19*(1), 53–61.

Marzano, R. J., Pickering, D., & Pollock, J. E. (2001). *Classroom instruction that works: Research-based strategies for increasing student achievement.* Association for Supervision and Curriculum Development.

Nicholson, T. D. (1961). A planetarium demonstration in the classroom. *Curator: The Museum Journal 4*(4), 295–303.

Noble, M. K. (1964). The planetarium and space science in the elementary school. *Science Education, 48*(1), 28–31.

Ortell, E.D. (1977). *The value of the planetarium as an instructional device.* Doctoral Dissertation. Nova University.

Reed, G. (1970). Is the planetarium a more effective teaching device than the combination of the classroom chalkboard and celestial globe? *School Science and Mathematics, 70*(6), 487–492.

Reed, G. (1972). An outline for the education of a K-12 planetarium teacher. *Journal of College Science Teaching, 1*(3), 51–52.

Reed, G. (1973). The planetarium versus the classroom—an inquiry into earlier implications. *School Science and Mathematics, 73*(7), 553–555.

Reed, G. (1976). Can first-graders be taught night and day in the planetarium? *School Science and Mathematics, 76*(7), 545–550.

Rice, V. (Spitz), & Towne, J. (2013). Who was Armand Spitz? *Planetarian 42*(4), 34–38.

Ridky, R. W. (1973). *A study of planetarium effectiveness on student achievement, perceptions and retention.* Doctoral Dissertation. Syracuse University.

Ridky, R. W. (1974). *A study of planetarium effectiveness on student achievement, perceptions and retention.* Paper presented at the National Association for Research in Science Teaching, Chicago, IL. Available online at: http://eric.ed.gov/?id=ED091207.

Ridky, R. W. (1975). The mystique effect of the planetarium. *School Science and Mathematics, 75* (6), 505–508.

Rischbieter, M. O, Ryan, J. M., & Carpenter, J. R. (1993). Use of microethnographic strategies to analyze some affective aspects of learning-cycle—based mini-courses in paleontology for teachers. *Journal of Geoscience Education 41*(3), 208–218.

Rosemergy, J. C. (1967). *An experimental study of the effectiveness of a planetarium in teaching selected astronomical phenomena to sixth-grade children.* Doctoral Dissertation. University of Michigan.

Schatz, D., & Friedman, A. (1976). Self-discovery in astronomy for the public. *Sky and Telescope, 52,* 254.

Slater et al. (2015). *Conducting astronomy education research: an astronomers' guide.* Hilo, Hawai'i: Pono Publishing.

Smith, B. A. (1966). *An experimental comparison of two techniques (planetarium lecture-demonstration and classroom lecture-demonstration) of teaching selected astronomical concepts to sixth grade students.* Ed.D. Dissertation. Arizona State University.

Sunal, D. W. (1973). *The Planetarium in Education: An Experimental Study of the Attainment of Perceived Goals.* Doctoral Dissertation. University of Michigan.

Tuttle, D. E. (1965). *Effects of the use of the planetarium upon the development of spatial concepts among selected sixth grade students in Elgin* (Master's thesis), Northern Illinois University.

Wright, D. L. C. (1968). *Effectiveness of the planetarium and different methods of its utilization in teaching astronomy.* Doctoral Dissertation. University of Nebraska.

Chapter 4
Learning Research in the Planetarium After 1990

The 50-year period after the 1930 opening of the first formal planetarium in Chicago hosting a towering Zeiss star projector might be considered the *golden age* of the planetarium. During this period, more than 1000 planetarium facilities were installed across the United States. Many of these planetariums were hosted at schools and colleges, and were much smaller than the mammoth ones hosted at cornerstone museums. These were often funded in part by the federal government as part of the 1958 US National Defense Education Act, which was created in the wake of Sputnik's launch at the very dawning of the Space Age when the US desperately needed to produce more mathematicians, scientists, and professors. This was a time in US history when a significant fraction of all school children visited planetariums to learn the constellations, observe the motion of the heavens, and be inspired to support and contribute to the US's efforts to become a leader in the Space Age (Marche 1999).

Paradigm Shifts

The last two decades of the 20th Century were characterized by tremendous changes to the United States educational system and by dizzying advances in off-the-shelf technology available at ever decreasing costs. Those changes caused a tsunami of impacts in the planetarium education world. Whereas the golden age of planetariums could be characterized as unrestrained growth and enthusiastic construction of new facilities, the current age is significantly and perhaps innumerably different. These changes to planetariums included, to name just a few: a dramatic shift from projecting the stars using internally illuminated star balls to using computer-driven video projectors; changing from live lectures delivered by an astronomy education expert to pre-recorded, multi-media productions; and a sea change in scientific content, from shows designed by artists illustrated by panoramas and single special effect projectors, as well as the ubiquitous constellations and

© The Author(s) 2017
T.F. Slater and C.B. Tatge, *Research on Teaching Astronomy in the Planetarium*, SpringerBriefs in Astronomy, DOI 10.1007/978-3-319-57202-4_4

planets current night sky shows, to delivering actual science data, images, and video highlighting current news about scientific research advances concerning mysterious astronomical objects observed with high-technology equipment; and large theaters experiencing unexpected financial stress because of the faltering US economy. Planetariums increasingly began to use their domes for revenue-producing laser shows, instead of the staple of educational night-time star tours, to garner sufficient revenue to make the financial books balance (viz., Reed 1994; Hitt 1999).

At the same time, the nature of planetarium educational research evolved quickly. For one, the once locally controlled K-12 school districts were more than ever becoming influenced at the federal level. Several influential groups were aggressively advocating for new national consensus guidelines for what astronomy should be taught in all K-12 schools in the form of national standards and curriculum frameworks (Adams and Slater 2000; Slater 2000). For another, the nature of publishing astronomy education work was changing. Whereas prior to the 1990s, when scholars completed their research dissertations, the scientific results of that work was only rarely published in refereed journals. Today, in contrast, research dissertations, along with professional meeting contributions, are usually considered to be in the *grey literature* of insufficient jurying (Slater 2015), and only peer reviewed journal articles in sufficiently high ranking journals count as scientific evidence. The end result of this cultural change is that if today's scholars of planetarium education only read peer reviewed journal articles, they might mistakenly conclude that little to no planetarium education research was done prior to 1990, which is demonstrably false.

Moreover, a significant emerging challenge to contemporary United States' planetarium educators has been how to deal with rapidly growing diversity among US K-12 students, many of which are demographically different than the planetarium educators. There are a wide variety of ways to classify which students are a *minority*. As one measure, consider the number of public school children who are non-Native English speakers, also known as English Language Learners (ELLs). This is a rapidly growing demographic across the US. During the 2012–2013 academic year, Alaska, California, Colorado, Nevada, New Mexico, Texas, and the District of Columbia had at least 10% of their students classified as ELL, with California having 22.8% of students being ELL (NCES 2015).

Furthermore, what was being considered as rigorous evidence in planetarium education research was evolving. The last part of the 20th Century saw the education research *paradigm wars*, which pitted the more quantitatively-driven, positivist educational researchers with their powerful statistical tools and numerical recipes against the more qualitatively-driven, interpretive educational researchers who used phenomenological interviews, social science anthropological observations, and self-reflections to make meaning out of transformative educational experiences (Gage 1989). The end result of these research methodological paradigm wars were not that one approach was better than the other, but instead that both quantitative and qualitative approaches had an important role in making sense out of the complexities of education. Furthermore, the type of data collected and published by planetarium education researchers became substantively more varied

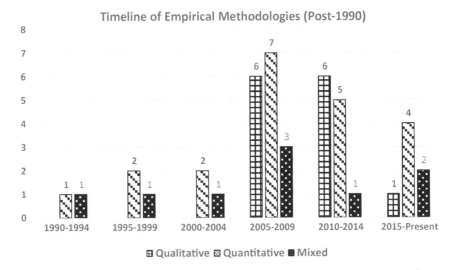

Fig. 4.1 Distribution of research methods in planetarium education research reports 1990 to present

than what was seen in the golden age, which only used quantitative methods (Fig. 4.1).

At the dawning of the 1990s, Twiest (1989) was working at the University of Georgia and visited the age-old question of whether students learn best in school classrooms or in the rich environment of the planetarium. He used a two-group, comparative pre-test posttest study design administering a self-developed multiple-choice knowledge (and attitude) survey to 423 4th, 5th, and 6th grade students (sample item stems are provided in Table 4.1). Uniquely, at the time, he disaggregated his data by gender. Upon analyzing his data, he found that students who learned in the classroom setting had statistically higher scores than those who learned in the planetarium, regardless of gender or grade level. These results were similar to work done decades earlier before planetarium education researchers finally understood that the planetarium needed to be used as an interactive class-room, rather than as a high-tech presentation facility (e.g., Reed 1972).

Portable Planetariums

About this time, portable planetariums were making their way into US school buildings. Initially, the most popular planetarium system widely available was the *STARLAB*. The *STARLAB* planetarium was invented by Philip Sadler and his class of middle school students in Lincoln, Massachusetts in about 1977. It was quickly mass produced and widely disseminated in the 1990s, often using federal grant

Table 4.1 Characteristics of golden and modern planetarium education ages

Golden age of planetarium education	Modern age of planetarium education
Mechanical star balls and 35 mm slide projectors	Computer-driven video projection
Rapid construction of planetariums	Portable planetariums contribute to portfolio
Concentric seating around star ball	Unidirectional seating
Live lectures	Turn-key multi-media presentations
Focus on moving celestial sphere	Focus on current scientific advances
Astronomy presentations	Interdisciplinary topics, converted IMAX movies, and music laser shows
Locally controlled school district curriculum	National, common core curriculum standards
Most students are native English speakers	Rapidly growing diversity of student demographics
Programs developed locally	Combination of commercial and NASA federal funding of programs, from free to expensive
Quantitative research methods	Quantitative, qualitative and mixed research methods
Dissertations, theses and meeting contributions dominate scholarly dissemination	Refereed journal articles & book chapters disseminated via internet dominate scholarship

award money in support of the purchase. There are more modern versions of portable planetariums today, but the original dome was a tarp-like sphere inflated by a conventional floor box fan and used a bright light bulb surrounded by an opaque plastic cylinder with holes to project stars. The *STARLAB* planetarium only cost 1/10th to 1/100th that of a permanent planetarium, and had the advantage of being transported to schools rather than having the children loaded onto an expensive school bus and taken to the planetarium (Cheney 1991; Manning 1996). The *STARLAB* was also widely heralded as a novel way to engage students in learning science and mathematics that would most certainly positively impact learning and achievement (viz., Hurd 1997).

A decade later after Twiest (1989), Meyer (2000) used Twiest's survey instruments to reproduce a study by Wright (1968); however, she might not have been aware of his work, as it was not cited in her dissertation. The study discussed the value of pre- and post-planetarium visit learning activities, not in a permanent brick and mortar planetarium installed in a building, but inside an inflated, portable *STARLAB* planetarium. She designed a multiple treatment, pre-test posttest study design to see if students who participated in various pre- and post-visit activities would increase the number of items they answered correctly. For a twenty item survey (shown in Table 4.1), she calculated posttest minus pre-test gain scores for three different treatment groups. One treatment group experienced hands-on activities before and after their planetarium visit, and had an average gain score of 3.14 questions correct (n = 56). A second treatment group used audio-visual activities before and after their visit, and had an average gain score of 2.97

questions correct (n = 60). A third treatment group completed text and reading based activities before and after the planetarium, and increased their scores with a gain of 3.72 questions correct (n = 65). Statistical analyses suggested that there were no significant differences among the various treatment groups, although the pretest to post-test score increases are statistically significant, if small in effect size. Regardless, the effect of the interventions were all quite small, which is consistent with what Wright (1968) found decades earlier: although pre- and post- field trip activities do appear to be important in many field trip contexts, it had not been shown to be true for visits to the planetarium.

Desktop Planetarium Software

The *STARLAB* was not the only less expensive option that planetarium education researchers were exploring to improve students' learning in astronomy. Desktop-style computers were becoming ever more ubiquitous in schools, and these computers were able to simulate planetariums. Baxter and Preece (2000) compared the learning of 48 5th and 6th grade students in the United Kingdom using an 11-item multiple choice pre- and posttest. Some of the students visited a planetarium, while others used the *Voyager II for Macintosh* computer simulation of a planetarium. The data that was collected showed no statistically significant differences between either of the treatments. However, they noted that female students had noticeably lower pretest scores than male students (albeit not statistically significant, and therefore could have happened by random chance). When considering the posttest, female students who learned in the planetarium had larger gains than their male counterparts; however, these results were not statistically significant.

In a purposeful effort to make seemingly complex science information presented in a modern planetarium more appealing to visitors, Fisher (1997) explored the use of humor in the planetarium to see if it increased the participants' levels of learning. Taking data from 495 adults, all over the age of 18, he presented participants with one of two versions of a pre-recorded, 15-minute planetarium show who then took a posttest. In the humorous-version of the show, a short quip-like humorous anecdote or pun-like joke was inserted into the show about every 90 s. In the non-humorous version of the show, these humorous inserts were removed. A total of 20 concepts were described in the show, 10 of which had humorous inserts and which alternated with the 10 non-humorous concepts. Table 4.2 lists the participant-supplied response test items. The questions shown here were taken directly from the show's script (Table 4.3).

The results showed that 245 adults viewing the humorous-version of the show scored 64.3% correct, whereas the 250 adults viewing the non-humorous-version of the show scored 68.0% correct. This difference showing that the non-humorous show is more effective is statistically significant ($p = 0.0202$), given the large number of subjects; however, the effect size is quite small. These results showing that extraneous information—in this case, humor—has negative impacts on

Table 4.2 Sample survey questions from Twiest (1989)

1. The ecliptic is the apparent path of the _____
2. Which planets orbit the Sun in less time than Earth?
3. How will the length of a year on an outer planet compare with a year on Earth?
4. Which object is often referred to as a morning or evening star?
5. Can the planet Venus ever be seen at midnight?
6. Do the planets move east-to-west or west-to-east against the background of stars?
7. How many constellations are there in the astronomical zodiac?
8. The constellation of Virgo is visible in the evening during what month?
9. What is the constellation along the ecliptic just east of Cancer?
10.If a planet was west of the Sun, at what time could you see it?

Table 4.3 Short Answer test questions used by Fisher (1997)

Form A
1. How does a planetarium project stars onto the ceiling?
2. What causes the sky to change color during a sunset?
3. Why do night-time objects appear to move across the sky?
4. Why is the Moon bright?
5. Three bright stars which form a triangle are named what?
6. We can find what sky picture by using the Big Dipper?
7. Who did Zeus send flying into the sky?
8. What are tiny pieces of rock or dust floating in space?
9. What night-time event occurs every year in August?
10. If you look in which direction you will see the Sun rise?
Form B
1. Why should you not look up at the Sun?
2. What is a group of stars that forms a picture in the sky?
3. City light reflected into the sky is called what?
4. The rings of Saturn are made up of what?
5. The Big Dipper has how many stars in it?
6. The Greeks thought the Big Dipper looked like what?
7. What is the only star that does not appear to move?
8. What causes objects to burn as they enter the atmosphere?
9. What causes a comet's tail to form?
10. What is the study of space and celestial bodies called?

learning is consistent with earlier education research reports by Wandersee (1982), more recent multimedia theory, and experimental research by Mayer et al. (2001), among others.

Hands-on Activities in the Dome

As part of ongoing efforts to make planetarium shows more effective at measurably increasing learning, Rusk (2003) studied how hands-on science activities heretofore part of the science classroom domain might enhance learning if done as part of the planetarium learning experience. He worked in a suburb of Dallas, Texas with high numbers of minority students (the Caucasian school population was less than 50%) and wanted to find ways to more successfully engage students' curiosity when they came to the planetarium. Using a large scale, two-group, comparison study, he first showed students an age-appropriate, pre-recorded planetarium program about phases of the moon (*Moonwitch*, written by Phil Groce for Bowen Productions). Following the show, half of the students completed a hands-on activity before taking the posttest while the other half of students took the posttest without the benefit of the post-visit activity. It is important to note that in a purposeful effort to avoid denying the other half of his students what he believed to be an important educational activity, the "control" group of students eventually went on to partic-ipate in the hands-on activity, after they completed their posttests. In this way, Risk (2003) made sure that all students had the opportunity to learn through the hands-on activity.

The post-show classroom activity was designed to help students better under-stand the cause of moon phases, a topic notoriously challenging for students to learn (Bailey and Slater 2005). Each student was given a 3" Styrofoam ball friction mounted on a 9" stick. Students worked through a teacher-directed activity and held the "moon" at arms' length and rotated themselves to re-create the "phases of the moon" resulting from a bright light mounted on one wall at eye-level. Students also recreated conditions for eclipses to occur, and not to occur.

After administering a 14-item posttest to about 1900 5th grade students that included both multiple-choice items and matching items he considered to be "difficult," Rusk (2003) found that students who completed the activity scored an average of 19.48% better than students who had not yet completed the activity. He interpreted this as evidence of the critical importance of using hands-on activities as part of instruction. To further explore the data Risk (2003) collected, the data was disaggregated based on the socio-economic demographics of the participating schools. His analysis showed that students from low socio-economic backgrounds benefited more from conducting the hands-on activity than other students. He was not, however, able to use a pre-test in his study design to isolate the impact of the planetarium show on learning; the planetarium program perhaps had little to no impact on students' understanding. His results do add considerable weight to the much earlier work of Giles (1981), and Edoff (1982) that best practices in classroom science teaching using active learning strategies grounded in sound educational theory are also effective when enacted in the planetarium environment with fidelity.

Five years later, Palmer (2007) also conducted a study on students' learning about the moon phases using the same *Moonwitch* planetarium program used by Risk (2003). In his study sample of 178 5th grade students, 41% were Hispanic, 34% were White, and 23% were African American. All 178 5th grade student-participants first took a common pretest then received classroom instruction on the moon's phases and eclipses. The students were then divided and a subset of 85 students attended the *Moonwitch* planetarium program, whereas a comparison group of 93 students attended a second 45-minute classroom instructional session. Finally, all students completed a posttest.

The pretest-posttest instrument used for the study included 10 multiple-choice items, a short student-supplied response essay section, and a manipulative section. The manipulative section required students to physically move and sequence eight pictures of the moon from the start of the lunar cycle to the end of the cycle. Summarized in Table 4.4, posttest minus pretest gain scores were calculated for all student participants. The statistical difference in the observed gain scores were significantly different for the two groups, with students who attended the planetarium program instead of equivalent class instructional time scoring significantly higher. When the data was disaggregated according to demographics, these gain scores were statistically significant for all subgroups. The African American students attending the planetarium program made dramatically larger gains in pretest to posttest scores compared to other demographic subgroups. This work suggests that earlier positive learning gains by Rusk (2003) might have been due to some inherent effectiveness of the *Moonwitch* planetarium show, and that his hands-on activity using balls mounted on sticks simply enhanced students' understanding even further, despite the lack of a sound educational theory to support this supposition a priori. One perhaps important aspect of *Moonwitch* is that *Moonwitch* utilizes a coherent and personalized story-line to describe the causes of moon phases and eclipses. Very recent theoretical work by Slater (2014) suggests that story-oriented planetarium shows could be more consistently effective than impersonal, information-download planetarium presentations, but this theoretical work is still premature.

Table 4.4 Palmer's (2007) two group comparison study design

Treatment	Phase I	Phase II	Phase III	Phase IV	Gain scores
1 (n = 85)	Pretest	Conventional classroom instruction	Attended *Moonwitch* planetarium program	Posttest	6.85 (SD = 9.20)
2 (n = 93)	Pretest	Conventional classroom instruction	45-min of additional classroom instruction	Posttest	3.64 (SD6.73)

Multiple Intelligences

Gardner (2000) proposed a wildly popular theory of learning that many classroom teachers quickly adapted into their teaching approaches. Called the *theory of multiple intelligences*, Gardner argued that all learners had domain-specific aptitudes, or intelligences, that they brought into learning situations that, if appropriately leveraged by a talented teacher, could improve student learning. He listed these intelligences as musical-rhythmic and harmonic; visual-spatial; verbal-linguistic; logical-mathematical; bodily-kinesthetic; interpersonal; intrapersonal; naturalistic; and existential, and advocated that traditional intelligence quotient IQ tests only measured a subset of these domains in a highly constrained way (*c.f.*, Willingham 2004).

Sarrazine (2005) revisited the much earlier work by Wright (1968) to study the educational impact of pre-planetarium learning activities and post-planetarium learning activities, but using a contemporary learning theory. Focused on improving students' understanding of the moon at its phases, her pre- and post-planetarium visit activities were created to be consistent with and leverage the cognitive notions of Gardner's (2000) multiple intelligences. She created two different, 50-minute classroom activity sequences that were constructed with activities designed to utilize seven of the intelligences: logical-mathematical by measuring angles with a protractor, spatial by using the models to simulate the phases of the moon, bodily-kinesthetic by having to physically move themselves into positions to see the changes in phase on the moon models, linguistic by reading and listening to directions and filling out data sheets, naturalist by studying the moon, and interpersonal by discussing their observations and results with their partners and table mates (Table 4.5).

Moreover, Sarrazine (2005) designed a special multiple intelligences planetarium program that purposefully used six multiple intelligences: (a) logical-mathematical by measuring angles between the sun and moon, (b) spatial by using the three dimensional planetarium setting to simulate the night sky and moon phases, (c) linguistic by reading and listening to information about the moon and moon phases, (d) musical by listening to music and singing about moon phases, (e) naturalist by studying the night sky and the natural phenomenon of the moon and its phases, and (f) bodily-kinesthetic by dancing and measuring angles with arms and hands.

Table 4.5 Sarrazine's (2005) six treatment groups	Group	Treatment sequence		
	1.		Planetarium program	
	2.	Activity A	Planetarium program	
	3.		Planetarium program	Activity A
	4.	Activity A	Planetarium program	Activity B
	5.	Activity B	Planetarium program	
	6.		Planetarium program	Activity B

To measure 483 6th–8th grade (10–15 year old) students' improvement in knowledge about lunar phases, she used a student-supplied response questionnaire and a 25-item multiple choice test in a multiple group comparative pretest post-test study design. All participating student groups had pretest to posttest gain scores that were statistically significant as a result of the various treatments. However, there were no statistically significant differences between the various treatment groups. Overall, student groups who participated in one or more of these particular classroom activities did not score better than the group that was part of the planetarium only treatment. This particular study design could not reveal if the classroom activities were unnecessary because her innovative planetarium program design was highly effective or because the pre- and post-visit classroom activities were largely irrelevant.

Leveraging a growing body of learning sciences research on how students' thinking evolves over time (Schweingruber et al. 2007, pp. 213–250), Plummer (2006) designed a study to describe what children know about celestial motion and to investigate the degree to which young children could learn these ideas in a planetarium-like environment. She conducted phenomenological interviews with 1st, 3rd, and 8th grade students (n = 60) where she could record both their verbal answers and observe any gestures students made when they described their thinking. Interviews revealed that most students, regardless of grade level, could accurately describe common motions in the sky.

Plummer (2009a, b) went on to use kinesthetic teaching techniques of gesturing and embodied cognition (Cook et al. 2008; Slater et al. 2008) to teach 1st and 2nd graders in Michigan about celestial motion. Data from 63 pre- and post-instructional interviews showed substantive improvement in understanding celestial motion and, at times, even surpassed the pre-instructional knowledge of 8th grade students. Her pre-instructional interviews lasted 12.0 min on average and her post-instructional interviews averaged 10.5 min. Her work clearly demonstrated the immense durable, educational value of having students physically using their hands to point to objects in the sky and then trace the motion of those objects across the sky.

Such an instructional approach of having students "dance the sky" seems to have durable impacts on student understanding. There is a need for robust instructional sequences that demonstrably help students develop meaningful understanding of complex phenomena, especially if such phenoema substantively tax students' cognitive abilities involving spatial reasoning tasks that are critical to the success of planetarium education (Plummer 2014). Small and Plummer (2014) examined the long-term durability of students' understanding of the Moon and lunar phases more than a year after initial instruction in both the classroom and in the planetarium. During the planetarium portion of instruction, students observed a character drawing observation in a sketchbook and were actively engaged in instruction by kinesthetically pointing to the Moon's position as it rose, moved across the sky, and set during several different lunar phases. Eleven 1st grade children were interviewed before instruction as a pretest, immediately after instruction, and then a third time more than one year after instruction as a delayed posttest. During the interview,

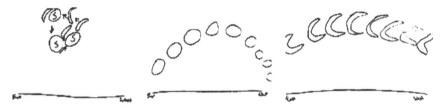

Fig. 4.2 Example of a 1st grade student's pre-instructional sketch of lunar phases (*left*); immediately post instruction (*center*); and one-year after instruction (*right*) adapted from Small and Plummer (2014)

students were asked to illustrate how the moon appears throughouth the day/night on a page that already had a horizon labeled with the cardinal points of East and West already identified (Fig. 4.2).

Using a Wilcoxon signed rank statistical test, data from demonstration interviews by Small and Plummer (2014) showed that students significantly improved from pre-instruction to post-instruction. Seven students retained their knowledge for more than a year, whereas four students regressed back to their immediate, post-instruction understanding level. Student participants were also asked how much time it would take to observe the entire lunar cycle. Prior to instruction, 36% stated it would take about one month to see all the phases. Immediately after instruction, the number of students stating it would take one month to see all the phases significantly increased to 73%, which decreased somewhat to 55% one year later. These results are interpreted as the inherent value of having students embody the patterns of motion that can be observed in the planetarium, and the long-term lasting positive effects of instructing students in this way.

Invoking Education Research in Planetariums

As extramural funding in the United States for planetarium programming projects—but not construction—reached its pinnacle in the mid-2000s. At the same time, the need for planetarium education researchers to show the value of federal funding in learning enhancement and the field of planetarium education also increased. Sumners (2008), from the Houston Museum of Natural Science, published a multi-site, pretest-posttest evaluative study of planetarium programming in high minority population centers. The three study populations of elementary grade students were: 221 inner city, Houston 7th grade students (who visited a full-dome video facility), 83% were Hispanic, 96% in poverty; 201 Navajo students, living in New Mexico (who watched the full-dome video in a portable planetarium, 92% of which were Native American; and a predominantly Hispanic group of 54 5th grade students (who also watched the full-dome video in a portable planetarium), many of which were fluent Spanish speakers, but reportedly not strong Spanish readers. In all cases, the test was given in English, although it was offered in Spanish as an option that was declined.

Student-participants took a 17-item multiple-choice test before and after watching a full-dome, 22-minute length show—not counting pauses in the program where the attending planetarium educator stops the program to interact with students—called *Earth's Wild Ride*. Like *Moonwitch* described above, this program uses an imaginative story line centered on the notion of children who were born on the Moon learning about the distant Earth from their wise grandfather. The child characters ask questions of their grandfather using children's natural language and learn about the Moon and Earth through the lens of comparative planetology. They also compare the starkly different conditions on the two surfaces. Unlike *Moonwitch*, *Earth's Wild Ride* has places where the program presentation is stopped to provide the audience opportunities to interact with the planetarium educator on hand to talk about what is occurring in the program. Students were also encouraged to handle and compare Mammoth and Tyrannosaurus rex teeth; igneous and sedimentary rocks; and stony and iron meteorites.

Sumner's (2008) most commonly cited result is that the planetarium program resulted in a statistically significant pretest to posttest gain in student scores. This illustrates the learning power of a short, but immersive, program that uses both a story-line children can identify with an interacting planetarium educator at various points throughout the show. Students improved their multiple-choice survey score from an average of about 53% correct to about 71% correct, which was statistically significant.

However, some of the most interesting results came from matching the program's variable teaching modality to specific items on the knowledge survey. Sumner (2008) and colleagues categorized how ideas were presented in the planetarium program as hearing, seeing, observing discussion between characters, or experiencing or a combination of modalities. They then mapped these learning modalities onto the pretest posttest knowledge survey to assess which learning modalities were most effective at helping students improve on each individual test item. Their analysis clearly showed that students improved their scores on specific conceptual test items when they were observing a discussion about the concept between two characters in the show. When concepts presented were isolated as a hearing or seeing experience only, these learning modalities had the lowest impact on student learning gains as measured by the survey instrument. Perhaps unsurprising, students made the greatest gains when a concept was presented as a combination of all three modalities. Taken together, Sumner's (2008) results add weight to the notion that the most effective planetarium programs use a comprehendible story line (viz., Slater 2014) or *cultural wrapper* (viz., Heimlich et al. 2010), and that purposeful discussion between or among characters emphasizes the scientific concepts being taught.

Working to try to untangle what stories capture students' attention and simultaneously helps them learn without being distracted, Gillette (2013) designed a study to determine if learning can be enhanced in an inflatable, portable planetarium if careful attention was paid to recent results of cognitive psychology and learning was applied to multi-media learning theory (Mayer 2008, 2009). He designed a two-group comparison study of two different planetarium programs to determine if extraneous but attention-garnering seductive details had positive or negative influences on student learning. As such, his study was similar to an earlier study by Fisher (1997), who found that including humor unnecessarily has negative impacts on students' learning.

Unintended Distractions

One might naturally assume that if curious or interesting pieces of information are added to an otherwise straightforward information-laden presentation, that learners might pay more attention and be able to learn more from the presentation. In contexts outside of astronomy, Mayer (2009) and colleagues repeatedly found that in carefully controlled psychology laboratory experiments, that nearly everything not directly related to the information at hand was cognitively distracting and resulted in lower levels of learning measured. His massive collection of systematic experimental studies included verifying that unnecessary use of colors, changing of font styles and sizes, and extraneous images all negatively impacted learning. Simultaneous recommendations from within the planetarium education community were directly in conflict with Mayer's work. For example, Croft (2008) clearly advocated that the most effective planetarium shows *must*: (a) be an immersive dome, (b) incorporate music, (c) take the audience on a journey, (d) include live presentations, and (e) be a peaceful, relaxed environment. Some of these ideas are reminiscent of decades' earlier work where the unusual learning environment of the planetarium theater itself served as a distracting impediment to learning (Ridky 1975) (Table 4.6).

Gillette (2013) divided 56 5th grade students from southern California into two comparison groups from a diversely populated school district that had an ethnicity of 48% White, 30% Hispanic or Latino, and 8% African American. On average, the two groups were statistically equivalent groups on the pretest. The test included

Table 4.6 Sumner's (2008) evidence of effectiveness of different learning modalities

Learning modality	Improvement on posttest (%)
Participants hear the concept described by narrator or an isolated character	40
Participants see the concept displayed graphically	39
Participants see and hear two or more story characters discuss a scientific concept	49
More than one learning modality employed	40
More than two learning modalities employed	59

Table 4.7 Sample test items from Gillette (2013)

Item type	Sample test item
Transfer (trouble-shooting)	On a dark moonless night far from any bright lights, how do the stars appear to be spread across the sky?
Transfer (what-if)	Imagine Earth had no air, rain, or clouds. What would the temperatures be like during the night?
Memory-Recall	What is the largest source of heat for the surface of Earth?

both memory-recall questions and transfer questions, a sample of which is provided in Table 4.7.

The lesson given to students in the traditional group was approximately 34 min in length and contained five topics: overview of the night sky, an explanation of the seasons, examples and diagrams of solar and lunar eclipses, a grand tour of the solar system, concluding with a depiction of the lunar cycle. The lesson presented to the experimental treatment group was essentially the same, but included 53 additional images (such as topic-related science fiction movie posters) which added an additional three minutes of instruction. On average, one of these additional images interrupted the lesson every 40 s and script deviations were experienced every 78 s. These interruptions were at a faster pace than Fisher's (1997) insertion of humor every 90 s.

At the end of the study, students from both treatment groups took the posttest. The treatment group that received the program without extraneous images, which Gillette (2013) called seductive details, increased their pretest average from 43% correct to 55% correct on the posttest. On the other hand, the treatment group that received the program with extraneous images only increased their pretest score of 42% correct to 47% correct on the posttest. The differences between the posttest minus pretest gain scores for the two groups are statistically significant, and the inclusion of seductive details in the program might have increased student interest, but negatively impacted their learning gains. These results are consistent with predictions from multimedia theory and experiment (Mayer 2008, 2009), and with earlier research by Fisher (1997).

The Field Trip

The vast majority of school-oriented programs given by planetariums in the US are targeted at the elementary school level. There are a variety of reasons for this, but the overarching reason is that in the United States, elementary students receive nearly all their instruction for all subjects from a single teacher in a single classroom. This is in contrast to junior high, middle school, and high school students, who take subject-specific courses from subject-specific teachers. In such a situation, a class of elementary level students is together all day long whereas secondary level students are scattered across the school, rarely forming a cohesive cohort. The result is that day-long field trips are much easier to arrange for a coherent cadre of elementary level students than for secondary students, who would miss their other classes and other subjects if they went on a field-trip to the planetarium. A secondary reason is that many elementary teachers lack confidence in teaching astronomy, and hope that a field-trip to the planetarium will provide much of the needed quality instruction in astronomy their students require according to school district guidelines (Slater 1993; Slater et al. 1996; Slater et al. 1999).

Planetariums that are installed in museums not only need to be able to serve large numbers of elementary-level school students, but they must also be able to serve families, sometimes with pre-school aged children. Prior evaluation work by Miller and Daguang (2011) suggested that clinical interviews with parents sitting with their children in family groups were one of a limited number of reasonably effective ways to assess which concepts pre-school students had learned. Parents were positioned to best evaluate and summarize the personal questions and conversations they had with their young children to gauge both children's interest and understanding.

Petrie (2013) designed a scholarly study to better understand why families with pre-school students attended planetariums and what parents thought their pre-school aged children were learning by attending a planetarium show. She conducted brief surveys and short interviews with children and their parents in family groups after they watched an early-childhood oriented planetarium show which focused on the following three concepts: The Moon is visible during the day; The sky contains the Sun, Moon, and stars, which many people understand by telling stories; and Astronauts move differently because there is less gravity on the Moon. She collected 57 questionnaires and conducted 12 interviews. The most common response from parents about what they thought their child (or children) learned from the show was that there was one Moon, and one Sun. The second most common response from parents was that the Sun and Moon can be in the sky at the same time, which is consistently a difficult idea for many US school children. When children specifically were interviewed, about 50% readily related that the Moon could be seen in the daytime. Although not directly related to planetarium education research on learning specifically, the primary finding from her study was that the parents' main motivation for attending the planetarium is that they are interested or want to stimulate interest in astronomy, *not* because the child has a prior interest.

Research in Museum Settings

Given that many planetariums are in museum settings, Schmoll (2013), a graduate student at the University of Michigan, wanted to explore the aspects of learning in planetariums when combining formal classroom and informal museum settings. She designed a multi-modal study to test a comprehensive curriculum on celestial motion that featured integrated instructional time in both the museum planetarium and in formal classroom learning environments.

This study was conducted in a 5th grade classroom at a public elementary school in southeast Michigan and a digital planetarium at a local natural history museum. The 29 student-participants had similar demographics to the larger school district's socioeconomically affluent population: 95% Caucasian, 2% Asian/Pacific Islander, 1% Black, 1% American Indian/Alaskan Native, and 1% Hispanic. A 15-day astronomy unit was constructed using strict adherence to theoretical structure of S. M.I.L.E.S. guidelines for integrating formal classroom learning environments and informal learning environments enthusiastically advocated by Griffin (1998). The unit included pre- and post-planetarium activities in which the planetarium field trip occurred in the middle of the first day of the unit. She also devised a sky observations recording sheet that students could use repeatedly throughout the unit (Fig. 4.3).

Using a combination of knowledge surveys, semi-structured interviews, individual case studies, attitude surveys, and analysis of student artifacts, she found that most students could describe the daily motion of the Sun, but were unable to describe its seasonal changes with respect to rising and setting. Similarly, most students were able to describe the length of the lunar cycle, but a number of students had great difficulty in describing the Moon's motion relative to the Sun, despite that it was emphasized during instruction.

Although her study design did not allow for her to isolate the effectiveness of the planetarium as an educational tool from the extensive pre- and post-classroom learning activities, her results were consistent with findings in other studies. Wright (1968) and Sarrazine (2005) showed that pre- and post-planetarium activities in and of themselves do not particularly enhance the learning that occurred in the

Fig. 4.3 Planetarium sky observations recording sheet (adapted from Schmoll 2013, p. 66)

planetarium itself, unless the classroom activities are rich in robustly theoretically driven, research-confirmed activities, such as those using advance organizers and clustering (Giles 1981).

Unraveling the Measured Impact

Difficulties with isolating the impact of planetarium education efforts when embroiled in a larger educational experience is consistent with earlier, multi-modal work in South Africa, where Lelliott (2007, 2010) conducted a massive, mixed-methods study of astronomy learning in a variety of instructional environments, a portion of which included surveying and interviewing students who visited the Johannesburg Planetarium. A major proportion of the results from his work come from evaluating surveys, interviews, and graphic conceptual maps drawn by students. Because students in his study had other learning experiences in concert with visiting the planetarium where they did learning activities both inside and outside the theater, it is difficult to isolate the impact of learning inside the planetarium theater specifically. Regardless, he reports that students were able to provide evidence of substantive learning about the Sun and solar system, as well as about the nature of stars, although less so than knowledge of the Sun and solar system. He suggests that this difference is likely due to the immediate relevance of the Sun in students' daily lives, making their understanding about the Sun more readily motivating than about knowledge of the distant stars. This is somewhat in conflict with students' affective descriptions that the planetarium presentations students most enjoyed were those involving stars and how the stars spin.

In total, his research adds considerable weight to the effectiveness of students learning astronomy using a robust combination of learning activities that leverage best practices in classroom science teaching, both inside the planetarium and outside. In complex studies like these, the planetarium learning experience becomes a confounding variable rather than a clearly identifiable influence on student learning and achievement.

A Turkish Study

In 2013, Turkish planetarium education researchers Türk and Kalkan (2015) used a conventional pretest–posttest control group, quasi-experimental study design to compare how the planetarium compared to conventional classroom learning. The authors classify this study as quasi-experimental because there are no random assignments of subjects to the two treatment groups. Although similar studies had been repeatedly conducted across the planetarium education research landscape in the United States and Europe, no studies had been conducted in Turkey.

This study used a 14-item astronomy knowledge survey based in large part on items created by Zeilik et al. (1998) and translated into Turkish. For the 120 students in the non-planetarium treatment group, there was no significant change in their pretest to posttest scores from 33.87% correct on the pretest to 33.86% on the posttest. In stark contrast, the average score for the 120 students in the planetarium treatment group increased significantly from a pretest score of 13.18% correct to 46.68% correct on the posttest. The most common student misconceptions mitigated by the Turkish planetarium show were those involving time zones; the Moon's phases during eclipses; Seasons, the noontime Sun's altitude; distances to the stars compared to the Moon and planets. As a result of this study, the authors called on Turkish schools to extend their efforts to visit nearby planetariums in service to improving students' knowledge of astronomy.

The Dome Versus Desktop Planetarium

As Reed (1994) predicted decades ago (Fig. 4.4), the overlapping interactions between full-dome immersive digital planetarium programs for teaching and desktop-computer-based planetarium simulators for teaching is becoming ever

Like it? It's your new planetarium!

Fig. 4.4 Reprinted with permission from *Planetarian*, Vol 23, #1, March 1994. © 1994 international planetarium society. Dr. George Reed, Artist

more conflated as technological advances in both domains accelerate. Zimmerman et al. (2014) designed an evaluation study to compare short- and long-duration learning gains of students who learned in a portable planetarium with a digital projection system to learning gains of students who used a flat-screen planetarium simulation on a desktop-computer.

After taking a 14-item multiple-choice pretest, 93 6th-8th grade middle school students viewed *We Choose Space* in a portable dome, while 107 middle school students viewed the same program on a desktop computer before completing a posttest and, eventually, a 6-week delayed posttest. Using paired *t*-tests and independent *t*-tests, students in both treatment groups made statistically significant and equivalent gains in achievement from pretest to posttest. Both groups had statistically equivalent pretest scores, and could be considered to be equivalent samples. However, when retested 6 weeks later as a delayed posttest, the students who experienced the program in the planetarium retained their gain scores, but students who experienced the program on a computer screen regressed most of their gains and their scores were statistically indistinguishable from their pretest scores. These results add weight to the notion that a passive presentation on a computer screen would not have the same long term effects that other instructional approaches using best practices of instruction might have.

Impact of Color in Cosmology

Further pushing the planetarium education research envelope on how computer-driven digital projections could be used to better teach students, Buck (2013, 2014) explored how scientific visualizations might best support student learning in cosmology in a large, urban planetarium facility. One of her guiding study questions was: Can mediational cues like color affect the way learners interpret the content in a cosmology visualization that might be used in a large, urban planetarium and how do color choices affect learning?

As a first step toward developing effective visualizations for use in the planetarium, Buck (2013) studied how 122 college students enrolled in an introductory astronomy survey course for non-science majoring undergraduates used color to interpret a scientific visualization. She created a 16-item content and demographic survey that had both multiple-choice items and student-supplied response items. Her content questions focused on determining how student participants interpreted an image showing formation of dark matter structure of the universe (Fig. 4.5). Participants were divided into three groups who were each shown the same image in one of three different color pallets and asked to choose the color corresponding to (i) dark matter, (ii) stars, (iii) empty space, and (iv) hydrogen gas, or to indicate that the thing in question was not visible in the visualization.

Using a Chi-square analysis, she found that the 33 student participants who saw the version that used white to indicate dark matter and blue to indicate empty space, were almost four times more likely to misidentify dark matter in the visualization

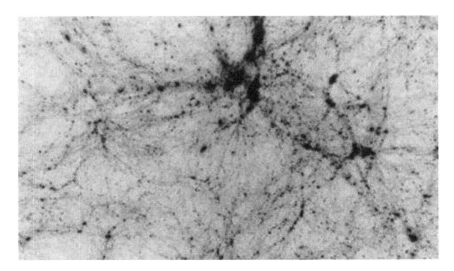

Fig. 4.5 Image showing formation of dark matter structure of the universe (from Buck 2013)

than those 58 student participants who saw a version where dark matter was indicated by a color that was darker than the background. Her conclusion was that the color chosen to represent dark matter had a statistically significant effect on students' interpretation of the visualization, and that planetarium educators and scientist-artists would benefit greatly by having interpretation data in making decisions about the color scheme used in order to improve understanding.

The 3D Stereoscopic View

The next major advances in planetarium education visualization might well not come from just appropriate color choice in visualizations, but instead from advances in projecting stereoscopically onto the planetarium dome. Price et al. (2015) have spent considerable effort, supported by extramural funding from the National Science Foundation and other entities comparing long-duration learning for 3-dimensional stereoscopic films displayed in large digital planetariums. Using quasi-experimental comparison field-tests to study the educational impact of various visualizations, 498 adult participants viewed either (i) a brief, conventional 2-dimensional simulation movie of the shape of the Milky Way galaxy, or (ii) a stereoscopic formatted 3-dimensional simulation movie of the same object. The study was quasi-experimental because adult participants were not randomly placed into experimental treatment groups.

What was educationally unique about the stereoscopic formatted 3-dimensional simulation movie illustrating the shape of the Milky Way galaxy was that design principles were derived from the learning sciences literature on spatial cognition and cognitive load. This had the express purpose of lowering the audiences' extraneous cognitive load and adhering to the principles in Mayer's (2008) multi-media theory (Price et al. 2015). The educational impact on adults viewing either the conventional 2-D or the novel 3-D stereoscopic films was compared using posttests and delayed posttests 6 months after initial viewing. A repeated analysis of covariances (ANCOVAs) statistical test with demographic and spatial visualization ability measures as covariates showed no differences in the short-term posttest learning gains between the 2D and stereoscopic treatments. However, the 6 month delayed posttest did show statistically significant differences with the stereoscopic treatment group having more durable learning gains. These results are motivating an international effort to provide stereoscopic visualizations at numerous science visitors' centers, including astronomy education efforts at facilities in Japan and Peru (Shiomi et al. (2014).

Clearly, the visualization possibilities that accompany digital planetarium projections are driving planetarium education research to measure the extent to which students can be taught challenging topics in astronomy education, such as the cause of Earth's seasons. Yu et al. (2015) divided 810 non-science majoring undergraduates enrolled in introductory astronomy survey courses into three distinct treatment groups: *Group I* were not provided with visualizations; *Group II* were shown visualizations on a flat screen projected at the front of the classroom; and *Group III* experienced instruction using full-dome immersive visualizations in the planetarium of the same visualizations *Group II* saw on a flat screen in a classroom. Using multiple-comparison groups with a pretest posttest study design to compare the posttest minus the pretest gain scores for each group, statistical analysis showed that the group who experienced visualizations in the planetarium had significantly greater gains than either of the two groups. The different gains for the college student treatment groups were: *Group I* gain score $\cong$ 4.4% more correct; *Group II* gain score $\cong$ 6.7% more correct; and *Group III* gain score $\cong$ 17.4% more correct. This result is consistent with similar, recent studies on middle school children (Chastenay 2016) and elementary school children (Carsten-Conner et al. 2015).

More Case Studies

The effects of digital planetarium programs on elementary students are smaller than on college students and adults, but still statistically significant. Carsten-Conner et al. (2015) studied 108 4th grade students in rural Alaska and had the opportunity to measure student learning in a portable planetarium using a digital, full dome projector. The impact of the program was measured using a pretest and a posttest, as well as think-aloud interviews with 10 students to support interpretation of the results. The program targeted four specific learning objectives: (1) the apparent movements of objects in the night sky are observable and due to Earth's movements; (2) stars appear to rotate around Polaris, the North Star; (3) the tilt of the earth on its axis causes the season and impacts the length of the day; and (4) like the

Earth, other planets orbit our Sun. The pretest to posttest gains showed statistically significant gains in student understanding, albeit with relatively small effect sizes.

Plummer et al. (2014) conducted pre- and post-instructional demonstration interviews with 99 3rd grade students who participated in one of four instructional treatments intended to help students better understand the apparent celestial motion of the Sun, Moon, and stars across the sky as a result of Earth's rotation. This was done in order to determine which instructional approaches produced positive learning gains. The four treatment groups compared were: (i) a space-based perspective taught in a conventional classroom; (ii) an Earth-based perspective as taught in a planetarium using kinesthetic teaching strategies of following sky motions by physically pointing at moving objects; (iii) classroom-based instruction focused on constructing explanations for the Earth-based observations; and (iv) a combination of learning experiences in the planetarium plus constructing explanations in the classroom. The student body demographics were representative of the larger school district within which the study population was drawn: 81.5% White, 2.1% Hispanic, 8.5% African-American, 4.8% Asian/Pacific Islander, and 3% Multi-racial students. The Kruskal-Wallis H test of statistical significance was used to confirm that students' pre-instructional knowledge was initially equivalent. This was done to minimize threats to internal validity due to non-randomization of student participants that were assigned to different treatment groups.

The Plummer and colleagues (2015) study was motivated from the perspective of trying to design effective instructional experiences to help students understand complex astronomical phenomena. Studying these types of phenomena often requires learners to use significant and cognitive taxing spatial reasoning skills. The study's data clearly showed that elementary students have the greatest learning gains when they engage in instruction that supports their ability to visualize Earth-based observations by engaging in multiple learning experiences that support each other: observe simulations and engage in guided gesturing, and participate in kinesthetic modeling.

In other words, it is not presence or lack of presence of a planetarium that makes a difference in student learning. Instead, it is what is done in each learning environment that works, especially when planetariums are used for what they are best at—showing celestial motion to actively engaged students. The digital planetarium can uniquely provide desperately needed cognitive support for students on difficult topics involving spatial reasoning that consume considerable cognitive resources on the part of learners.

The Digital Planetarium Impact

In total, what we are consistently seeing across the most recent decades in planetarium education research efforts is that the complex visualization capabilities of digital projection planetariums are beginning to show greater learning gains, particularly in the area of complex astrophysical concepts found in cosmology.

Decades earlier, systematic, comparative planetarium education research did not often show that the mechanical planetarium was any more successful of a teaching environment than a well-conducted classroom.

In perhaps seemingly archaic contrast to complex visualizations theory-based learning innovative learning strategies, the original goals of planetarium education research effort were to help students learn the night sky. Hintz et al. (2015) undertook the laborious task of cataloging what contemporary students know today about constellations upon entering college. Collecting pre-instructional data from 710 undergraduate students and 167 9th grade students visiting the planetarium, they found that students could identify only a few constellations and bright stars.

The most common constellations that entering university students could identify when pointed to by planetarium operators with laser pointers were, in order of frequency: Orion (70%), Ursa Major (47%), Draco (26%), Ursa Minor (22%), Canis Major (14%); Cassiopeia (11%) and Scorpius (10%), with all remaining constellations being identified by less than 10% of college students. In terms of stars, college students were able to identify Polaris (32%); Betelgeuse (24%); and Sirius (11%). For 9th grade students, results were similar to that of college students, but smaller: Orion (72%); Ursa Major (21%); Draco (12%) and Canis Major (9%). In terms of stars, 9th grade students were not asked to name particular stars in this study.

Although we can find no comparison data for students from six decades ago, we can assume that today's students rarely go outside or experience light polluted skies when they are outside. Planetarium education research efforts are still needed to measure how much students are learning. The effectiveness of teaching the science of astronomy beyond the recognition pattern of constellations is a fruitful direction for experiment and research.

References

Adams, J. P., & Slater, T. F. (2000). Astronomy in the national science education standards. *Journal of Geoscience Education, 48*(1), 39–45.

Bailey, J. M., & Slater, T. F. (2005). Resource letter AER-1: Astronomy education research. *American Journal of Physics, 73*(8), 677–685.

Baxter, J., & Preece, P. F. (2000). A comparison of dome and computer planetaria in the teaching of astronomy. *Research in Science & Technological Education, 18*(1), 63–69.

Buck, Z. (2013). The effect of color choice on learner interpretation of a cosmology visualization. *Astronomy Education Review, 12*(1).

Buck, Z. E. (2014). *Dynamic visualizations as tools for supporting cosmological literacy*. Ph.D. Dissertation, University of California—Santa Cruz.

Carsten-Conner, L. D., Larson, A. M., Arseneau, J., & Herrick, R. R. (2015). Elementary student knowledge gains in the digital portable planetarium. *Journal of Astronomy & Earth Sciences Education, 2*(2), 65–76.

Chastenay, P. (2016). From geocentrism to allocentrism: Teaching the phases of the moon in a digital full-dome planetarium. *Research in Science Education, 46*(1), 43–77.

Cheney, T. (1991). *A Comparison of the effectiveness of the portable STARLAB planetarium and the traditional classroom in the teaching of astronomical concepts.* Master's Thesis, State University of New York, College of Arts and Science, Professional Studies Division.

Cook, S. W., Mitchell, Z., & Goldin-Meadow, S. (2008). Gesturing makes learning last. *Cognition, 106*(2), 1047–1058.

Edoff, J. D. (1982). *An experimental study of the effectiveness of manipulative use in planetarium astronomy lessons for fifth and eighth grade students* (Ed.D. Dissertation). Wayne State University.

Fisher, M. S. (1997). The effect of humor on learning in a planetarium. *Science Education, 81*(6), 703–713.

Gagné, N. (1989). The paradigm wars and their aftermath: A "historical" sketch of research on teaching since 1989. *Teachers College Record, 91*(2), 135–150.

Gardner, H. E. (2000). *Intelligence reframed: Multiple intelligences.* Perseus Books Group.

Giles, T. W. (1981). *A comparison of effectiveness of advance organizers and clustering singly and in combination upon learning in the planetarium* (Doctoral Dissertation). Pennsylvania State University.

Gillette, S. (2013). *The effects of seductive details in an inflatable planetarium* (Doctoral Dissertation). Walden University.

Griffin, J. M. (1998). *School-museum integrated learning experiences in science: A learning journey* (Doctoral dissertation). University of Technology, Sydney.

Heimlich, J. E., Sickler, J., Yocco, V., & Storksdieck, M. (2010). Influence of immersion on visitor learning: Maya skies research report. *Edgewater, MD: Institute for Learning Innovation.* Retrieved from: http://www.informalscience.org/influence-immersion-visitor-learning-maya-skies-research-report

Hintz, E. G., Hintz, M. L., & Lawler, M. J. (2015). Prior knowledge base of constellations and bright stars among non-science majoring undergraduates and 14–15 year old students. *Journal of Astronomy & Earth Sciences Education, 2*(2), 115–134.

Hitt, R. J.,Jr., (1999). *A national survey of planetarium directors operating public-school-owned planetaria* (Doctoral Dissertation). Virginia Polytechnic Institute and State University.

Hurd, D. W. (1997). Novelty and it's relation to field trips. *Education, 118*(1), 29.

Lelliott, A. D. (2007). *Learning about Astronomy: A case study exploring how grade 7 and 8 students experience sites of informal learning in South Africa* (Doctoral Dissertation). University of the Witwatersrand, Johannesburg.

Lelliott, A. D. (2010). The concept of spatial scale in astronomy addressed by an informal learning environment. *African Journal of Research in Mathematics, Science and Technology Education, 14*(3), 20–33.

Manning, J. G. (1996). The role of planetariums in astronomy education. *Astronomy Education: Current Developments, Future Coordination, 89,* 80.

Marche, J. D. (1999). *Theaters of time and space: The American planetarium community, 1930–1970* (Doctoral Dissertation). Indiana University.

Mayer, R. E. (2008). Applying the science of learning: Evidence-based principles for the design of multimedia instruction. *American Psychologist, 63*(8), 760–769.

Mayer, R. E. (2009). *Multimedia* learning (2nd ed.). New York: Cambridge University Press.

Mayer, R. E., Heiser, J., & Lonn, S. (2001). Cognitive constraints on multimedia learning: When presenting more material results in less understanding. *Journal of Educational Psychology, 93* (1), 187.

Meyer, J. R. (2000). *The Effects of Three Types of Pre-And Post-Planetarium/STARLAB-Visit Instruction Methods On Astronomy Concepts And Attitudes Of Sixth Grade Students.* Ph.D. Dissertation, Kansas: University of Missouri.

Miller, J. D., & Daguang, Li. (2011). *The Impact of One World One Sky on Children's Interest and Learning about Astronomy.* Retrieved from http://www.informalscience.org/impact-one-world-one-sky-childrens-interest-and-learning-about-astronomy

NCES. (2015). *The condition of education 2015 (NCES 2015-144), English language learners.* Washington, DC: U.S. Department of Education, National Center for Education Statistics.

Palmer, J. C. (2007). *The efficacy of planetarium experiences to teach specific science concepts* (Doctoral Dissertation). Texas A&M University.

Petrie, K. B. (2013). *Early childhood learning in preschool planetarium programs* (Master's Thesis). University of Washington.

Plummer, J. D. (2006). *Students' Development of astronomy concepts across time* (Doctoral Dissertation). The University of Michigan.

Plummer, J. D. (2009a). Early elementary students' development of astronomy concepts in the planetarium. *Journal of Research in Science Teaching, 46*(2), 192–209.

Plummer, J. D. (2009b). A cross-age study of children's knowledge of apparent celestial motion. *International Journal of Science Education, 31*(12), 1571–1605.

Plummer, J. D. (2014). Spatial thinking as the dimension of progress in an astronomy learning progression. *Studies in Science Education, 50*(1), 1–45.

Plummer, J. D., Kocareli, A., & Slagle, C. (2014). Learning to explain astronomy across moving frames of reference: Exploring the role of classroom and planetarium-based instructional contexts. *International Journal of Science Education, 36*(7), 1083–1106.

Price, C. A., Lee, H. S., Plummer, J. D., Subbarao, M., & Wyatt, R. (2015a). Position paper on use of stereoscopy to support science learning: Ten years of research. *Journal of Astronomy & Earth Sciences Education, 2*(1), 17–26.

Price, C. A., Lee, H. S., Subbarao, M., Kasal, E., & Aguilera, J. (2015b). Comparing short-and long-term learning effects between stereoscopic and two-dimensional film at a planetarium. *Science Education, 99*(6), 1118–1142.

Reed, G. (1994). Who in the hell needs a planetarium? *Planetarian, 23*(1), 18–20.

Reed, G., & Campbell, J. R. (1972). A comparison of the effectiveness of the planetarium and the classroom chalkboard and celestial globe in the teaching of specific astronomical concepts. *School Science and Mathematics, 72*(5), 368–374.

Ridky, R. W. (1975). The mystique effect of the planetarium. *School Science and Mathematics, 75* (6), 505–508.

Rusk, J. (2003). Do science demonstrations in the planetarium enhance learning? *Planetarian, 32* (1), 5–8.

Sarrazine, A. R. (2005). *Addressing astronomy misconceptions and achieving national science standards utilizing aspects of mutiple intelligences theory in the classroom and the plaentarium.* Ph.D. Dissertation, Indiana University.

Schmoll, S. E. (2013). *Toward a framework for integrating planetarium and classroom learning* (Ph.D. Dissertation). University of Michigan.

Schweingruber, H. A., Duschl, R. A., & Shouse, A. W. (Eds.). (2007). Chapter 8—Learning progressions. In *Taking Science to School: Learning and Teaching Science in Grades K-8* (pp. 213–250). Washington, DC: National Academies Press.

Shiomi, N., Shoichi, I., Hidehiko, A., Mario, Z., José, I., Edwin, C., et al. (2014). Stereoscopic 3D projections with MITAKA: An important tool to get people interested in astronomy and space science in Peru. *Sun and Geosphere, 9*(1), 115–116.

Small, K. J., & Plummer, J. D. (2014). A longitudinal study of early elementary students' understanding of lunar phenomena after planetarium and classroom instruction. *Planetarian, 43*(4), 18–21.

Slater, S. J., Slater, T. F., & Morrow, C. A. (2008). The impact of a kinesthetic astronomy curriculum on the content knowledge of at-risk students. In *Proceedings of the National Association for Research in Science Teaching, Baltimore, MD.* https://narst.org/annualconference/annualprogram08_final.pdf

Slater, T. F. (1993). *The effectiveness of a constructivist epistemological approach to the astronomy education of elementary and middle level in-service teachers* (Ph.D. Dissertation). University of South Carolina.

Slater, T. F. (2000). K-12 astronomy benchmarks from project 2061. *The Physics Teacher, 38*(9), 538–540.

Slater, T. F. (2014). *How much louder do I need to turn up the soundtrack before they learn? A cognitive science perspective on memory in the planetarium.* Keynote Lecture at the Great

Lakes Planetarium Association Conference, October 30, 2014 in Muncie, Indiana. Retrieved from https://www.researchgate.net/publication/275947863 and from YouTube.com at: https://www.youtube.com/watch?v=Vu9MzFIwJMM

Slater, T. F. (2015). *Is The Best Astronomy Education Research 'Grey'?* Retrieved from AstroLearner blog: https://astronomyfacultylounge.wordpress.com/2015/09/24/is-the-best-astronomy-education-research-grey/

Slater, T. F., Carpenter, J. R., & Safko, J. L. (1996). Dynamics of a constructivist astronomy course for in-service teachers. *Journal of Geoscience Education, 44*(6), 523–528.

Slater, T. F., Safko, J. L., & Carpenter, J. R. (1999). Long-term attitude sustainability from a constructivist-based astronomy-for-teachers course. *Journal of Geoscience Education, 47*(4), 366–368.

Sumners, C., Reiff, P., & Weber, W. (2008). Learning in an immersive digital theater. *Advances in Space Research, 42*(11), 1848–1854.

Türk, C., & Kalkan, H. (2015). The effect of planetariums on teaching specific astronomy concepts. *Journal of Science Education and Technology, 24*(1), 1–15.

Twiest, M. G. (1989). *The Attitudinal and Cognitive Effects of Planetarium Integration In Teaching Selected Astronomical Concepts To Fourth, Fifth, and Sixth Grade Students* (Ph.D. Dissertation). University of Georgia.

Wandersee, J. H. (1982). Humor as a teaching strategy. *The American Biology Teacher, 44*, 212–218.

Wright, D. L. C. (1968). *Effectiveness of the planetarium and different methods of its utilization in teaching astronomy* (Doctoral Dissertation). University of Nebraska.

Willingham, D. T. (2004). Reframing the mind. *Education Next, 4*(3), 19–24.

Yu, K. C., Sahami, K., Sahami, V., & Sessions, L. C. (2015). Using a digital planetarium for teaching seasons to undergraduates. *Journal of Astronomy & Earth Sciences Education, 2*(1), 33–50.

Zeilik, M., Schau, C., & Mattern, N. (1998). Misconceptions and their change in university-level astronomy courses. *The Physics Teacher, 36*(2), 104–107.

Zimmerman, L., Spillane, S., Reiff, P., & Sumners, C. (2014). Comparison of student learning about space in immersive and computer environments. *Journal and Review of Astronomy Education and Outreach, 1*(1), A5–A20.

Chapter 5
Affective Domain Research in the Planetarium

Perhaps the most basic question planetarium education researchers are interested in can be simply summarized as, "Does visiting a planetarium make any difference?" In this context, the word *difference* can take on many different meanings. One set of differences might be from the cognitive domain of knowing, understanding, and applying: Does visiting the planetarium mean that people now know the names of stars, can find constellations, understand how planets move in the sky, or can better predict how high the noontime Sun will be in the summer so they can build better buildings? However, an alternative set of differences that could be just as important might be from the attitude domain of a persons' interests and values: Does visiting the planetarium mean that people now more intrinsically appreciate the beauty of the night sky, want to read books and newspaper articles about astronomy, begin to believe that astronomy should be a core class subject taught in every school, devote more of their personal time to looking at the stars, or even spend disposable income to travel across the planet and observe a solar eclipse?

Some of the differences might sit in the behavioral domain and even spur people into actions they would not otherwise have done if they had not been inside a planetarium: Does visiting the planetarium mean that some people are now compelled to pursue new learning or career opportunities toward a wider variety of non-astronomy scientific disciplines, contribute financially to the construction of a new museum, urge their city governments to enforce new dark-sky, low-lighting regulations for street lights, or are urged into action to buy a large backyard telescope so they can participate in the study of astronomy by contributing citizen scientist data to the professional scientific community? It is to these later questions that go beyond simply knowing constellations names and understanding motions of the celestial sphere that are a frequent study of planetarium education researchers. Whereas questions about what learners understand is known as the cognitive domain, questions about how learners feel about, and attend to are known as the affective domain.

The affective domain of planetarium education research is quite widespread in the number of issues one might conduct studies upon. Smith and Ragan (1999)

T.F. Slater and C.B. Tatge, *Research on Teaching Astronomy in the Planetarium*, SpringerBriefs in Astronomy, DOI 10.1007/978-3-319-57202-4_5

circumscribe the affective domain of education research to those educational outcomes that involve adoption and changing one's attitudes, motivation, and values. These are most often observed by planetarium education researchers looking for statements of opinions, beliefs, or an assessment of worth, regarding having a planetarium learning experience. The classic textbook most planetarium education researchers have on their shelf is *Assessing Affective Characteristics in the Schools* by Lorin Anderson and Sid Bourke (2000), although others are certainly widely used (sensu, Gable and Wolf 2012).

Attitude

The most commonly measured component of the affective domain is attitude. Attitudes vary widely in direction, amount, and intensity. Attitudes are usually characterized as positive, negative, uncertain, or mixed, and, depending on the researchers' agenda, are sometimes specified as desirable or undesirable. As one example, consider that a planetarium education researcher might be predisposed to hoping that planetarium attendees have positive attitudes toward taxpayer funding of astronomy research after visiting a planetarium show. As another example, consider that in 1958 the US Congress passed, and continued to fund for many years, the *National Defense Education Act*. Among other things, the Act supported the construction of hundreds of planetarium facilities in the hopes that students would be spurred on to pursue science and engineering careers. At the very least, students would be supportive of the US's taxpayer-funded engagement in the 1960s space race.

Do planetariums have a substantive role in addressing the attitude of its attendees? Longstanding luminaries in planetarium education research have often suggested that the most important role for the planetarium might be along the lines of enhancing variables of attitudes, values, and interest from the affective domain instead of improving memory and understanding variables from the cognitive domain. Without the benefit of systematic planetarium education research, Gardner (1964) and Vance (1964) both widely advocated that the appropriate use of the planetarium when used in conjunction with the teaching science in schools could provide that extra dimension needed to stimulate greater interest in science among students. These notions were echoed by planetarium education researchers a decade later. Based on his early research into student learning and attitudes, Reed (1973) concluded after his studies on both the cognitive domain and the affective domain that, "the real value of the planetarium may lie in the affective domain and that is where future research should be directed." His conclusion was echoed by Ridky (1974) working contemporaneously who enthusiastically believed that, "the effectiveness of the planetarium appears not to lie in facilitating content achievement, but rather in effecting attitudinal change."

Education and Emotion

Although some have naively described modern science endeavors as a dispassionate human enterprise, recent research by Sinatra et al. (2014) confirms that successfully learning science requires educators to address more than only the domain of knowledge, but that accounting for emotional components is critical to successful education. As but one example, consider the emotional blocks to learning that planetarium educators frequently encounter when teaching Big Bang Theory and billion-year long evolution of the expanding Universe. In the end, considering the affective domain is perhaps just as important to planetarium education researchers, if not more important, than addressing the cognitive domain of memory and conceptual understanding. A summary from Tatge (2016) and colleagues summarizing empirical dissertations and refereed publications in planetarium education research focusing on the affective domain is illustrated in Fig. 5.1.

As the United States scientific and engineering community edged closer to successfully landing humans on the Moon, Moore (1965) wondered if adults who attended planetarium shows in a Michigan community (Flint) might be different than those adults who did not. He specifically looked at measures of media participation, attitude differences, and vocabulary recognition between adult attendees and non-attendees with the goal of identifying what sorts of people attend planetariums and what benefits and inclinations they might derive from attending planetarium shows. Using a two-group, posttest only comparison study, he surveyed 107 adults who were attending a local, community education program. His

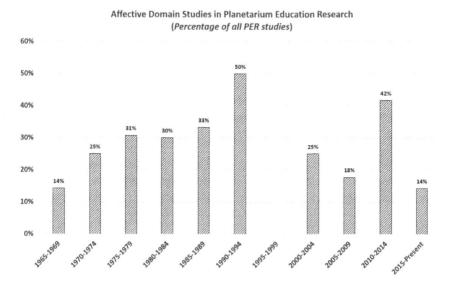

Fig. 5.1 Overview of empirical dissertations and refereed publications in planetarium education research focusing on the affective domain [adapted from Slater et al. (2016) and colleagues]

survey was a 12-item inventory sheet, and he compared responses among those that had voluntarily attended a planetarium show, and those that had not attended one.

Moore's (1965) results showed that planetarium show attendees and non-attendees responded differently. In terms of media habits, he found that adults who attended more often read newspapers whereas adults who did not, read books more often. His survey results also showed that adults who attend planetarium shows are more likely to travel to movie theaters and spend less time watching television, whereas planetarium non-attendees allocate more time each week listening to radio broadcasts. When considering attitudes toward taxpayer-supported space research expenditures, planetarium attendees appear to be more supportive than non-attendees. And finally, when looking quantitatively at survey responses, voluntarily visiting planetarium attendees used more "space vocabulary" terms in their responses than non-attendees. He concluded that the planetarium held great potential for impacting adults' behaviors with regards to media participation and enhancing adults' attitudes in supporting space-based research.

Audience Selection

In hindsight, there were two major limitations to Moore's (1965) study. The first is that his sample of 107 adults was acquired from people choosing to attend adult education classes. These individuals were already self-screened to a narrow sample of people who took the time and energy to pursue adult education classes, thus significantly constraining the ability to generalize his conclusions. The second major limitation is that he attributed the difference in planetarium-attending participants' attitudes towards space research expenditures. He also attributed his observed more frequent use of space-related words in attendees responses, as compared to non-planetarium attendees, to visiting the planetarium. Because he only conducted his study as a one-measurement, post-test only design, it is unknown if the planetarium attendance causes this increase of space-related words. It is equally likely that those who purposefully chose to attend a planetarium show already had considerable knowledge and interest in space-related topics.

This highlights one of the tremendous challenges planetarium education researchers face: The inherent difficulty of finding an appropriate and balanced study sample of humans who have and have not experienced planetarium shows among a vast ocean of possible astronomy-learning opportunities stretching far back into an individual's past. In other words, appropriate sampling is an enormously confounding and difficult-to-control experimental variable across all planetarium education research studies.

As schools were struggling to keep up with the rapid advances and ever-growing interest in astronomy due to the ongoing Moon landings in the late 1960s and early 1970s, Yee, Baer, and Holt (1968) undertook a two-group, multiple-measures, posttest and delayed posttest, control group comparison experimental study with randomized participants. The researchers surveyed 74 5th grade classrooms in

Wisconsin, where roughly one half of students visited the planetarium and one half did not. Additionally, a delayed posttest was given two to three weeks after visiting the planetarium. The results showed that there were no differences in students' attitudes toward astronomy, insofar as the instrument could measure.

Agents of Change

Considering the large financial investment a school, museum or community made in the construction of even a small planetarium, Jamison (1972) studied the extent to which the existence of a planetarium in a community—and the impact of a socially engaged and enthusiastic planetarium educator could serve as agents of change in enhancing social attitudes in schools and the broader community. By then, more than 1,000 planetarium facilities were in existence in the United State alone, constructed at an average cost of $2500 USD (1972) each. She designed a pretest-posttest comparative study surveying 228 community leaders (209 men and 19 women) using 12 biographical and demographical questions and 28 pretest and posttest attitudinal questions regarding social issues within the targeted communities. Seventy-two percent of respondents self-reported themselves as protestant, 69% identified with a preference toward the Republican political party, and most had never been to a planetarium before.

The pretest and posttest was given immediately before and after a 40-min planetarium presentation. Respondents were asked the extent to which they agreed or disagreed with a variety of statements. Examples of statements surveyed were: the community should offer more education classes and activities; the planetarium has a unique role to serve as a center for stimulation of the mind; and spiritual and moral values should also be taught during formal education process. A sample of her data is shown in Table 5.1.

Jamison (1972) clearly states that these are beliefs by community leaders, not experimentally verified truths which have been systematically measured. Nonetheless, the data consistently revealed a dramatic change, that was statistically significant, in self-reported attitudes about the value, importance, and need for a comprehensive community education program after a relatively short 40-min

Table 5.1 Example of survey results from Jamison (1972)

Statement 34. *"The Planetarium offers an unparalleled experience which aids clear understanding, promotes faster learning, and stimulates longer retention"*					
	Strongly Disagree	Disagree	Undecided	Agree	Strongly Agree
Pretest	n = 2 (0.9%)	n = 19 (8.3%)	n = 84 (36.8%)	n = 105 (46%)	n = 18 (7.9%)
Posttest	n = 0 (0.0%)	n = 22 (9.6%)	n = 33 (14.5%)	n = 116 (50.8%)	n = 57 (25%)

planetarium program to community leaders. This result is perhaps a bit startling as there is astoundingly little educational theory to explain a priori why a short planetarium program about the stars and the wonder of space would have such far reaching impacts. Yet, her results were impressive. Unfortunately, her work seemingly was not widely cited in the forthcoming literature, and her study was not extended nor her results confirmed.

Assessing Impact

By the early 1970s, more planetariums were constructed within school and college settings. Reed (1972, 1973) and colleagues adopted a posttest-only, control group comparative design with the randomized-groups to compare the impact on students' attitudes (and scientific knowledge) when using high-quality classroom instruction as compared to high-quality planetarium education when both settings are using education theory-driven, and experimentally verified best practices in science teaching. Reed (1970) measured clearly specified affective behavioral objectives using the planetarium as a teaching facility and compared it to a classroom instructional setting. His classroom-based instructional strategy was somewhat unique because he had students manipulate celestial spheres during instruction. His approach of making students more active learners in the process increased the quality of the non-planetarium instruction significantly, which he anticipated would also impact students' attitudes, values, and interests related to astronomy. After surveying 401 students who learned in the planetarium and 223 students who learned in a typical classroom at West Chester University, he could find no statistical difference between the two instructional interventions.

Undoubtedly, the planetarium has a certain degree of attractive charisma. At many colleges and universities, the number of students enrolling in an optional elective course is viewed as a rough proxy for the degree to which the course is valued by students. Reed (1975) wondered if simply having a planetarium available on campus would positively influence enrollment in a college astronomy courses.

He surveyed 112 undergraduate students enrolled in an astronomy course and 146 students enrolled in a general physical science course. Ninety-nine percent of students surveyed self-reported that they knew what a planetarium was, and 80% of all students reported that they had previously visited a planetarium. The only difference between the two groups were that 99% of astronomy students said they were aware of West Chester University's planetarium before they enrolled in the course, whereas only 78% of students enrolling in physical science were aware of their planetarium's existence. In either case, a substantial portion of the students were aware of the facilities existence and thus had great potential to impact students' course selection. Whereas 92% of physical science students reported that the existence of a planetarium had no influence on their election to take a physical science course, nearly 70% of college students deciding to enroll in astronomy said that the presence of the planetarium had some influence on their choice (Table 5.2).

Table 5.2 Influence of having a planetarium on enrolling in college astronomy course (Reed 1975)

Did knowing about the teaching of the course in the planetarium have any influence upon you selecting to take the college Astronomy course?	
Great influence	28%
Moderate influence	32%
Some influence	12%
No influence	28%

Although he apologizes saying, "the questionnaire survey is probably the weakest form of educational research; however … Hopefully more research will be conducted in the evaluation of the planetarium in the affective domain," his results strongly suggest that having a planetarium facility—and being sure students know about its existence—can help drive student enrollment in college astronomy courses.

Changing Attitudes

In the mid-1970s, Burnette (1976) conducted a surprisingly large-scale study to determine the role of the planetarium in changing the attitudes of 4th and 8th grade students in Columbia, Missouri. She used a two-group, pretest posttest study design to compare the impact of a newly implemented Earth-space science instructional program where one group (2748 students) did not attend the planetarium as part of the instructional sequence whereas the other group (2709 students) did visit the planetarium. She administered cognitive memory and knowledge surveys as well as attitude and interest surveys and compared the results using two-tailed t-tests for a difference between two independent means or differences between proportions, when appropriate.

Burnette's (1976) results regarding the affective changes in students were statistically significant for those visiting the planetarium. Of particular interest, when junior high school students were given a list of areas of science topics they were most interested in taking classes on, the choice of *astronomy* saw a statistically significant increase for 788 students surveyed who had visited the planetarium, compared to those 616 surveyed who had not, when compared to *biology*, *geology*, *chemistry* and *physics*. The percentage of students who reported that they "would enroll in an astronomy course if one were offered as an elective next year" was about 50% higher for students who visited the planetarium.

At the same time, when given a list of possible careers including: *medical*, *agriculture*, *space science*, *business*, *teaching*, *law*, and *construction*, the number of students choosing *space science* as a career also saw a statistically significant increase. However, these results applied only when comparing astronomy to other science disciplines. Her results did not show an overall increase in students' broader attitudes toward science in general for 4th graders and only minor enhancements for 8th graders. She recommended that further studies attempt to determine the optimal frequency of planetarium visits for maximum positive impact.

Electronic Response Devices Aid Surveys

Leveraging emerging technology for collecting information about planetarium visitors' attitudes in real-time during planetarium presentations, Gutsch (1978) created a comprehensive study to understand how attitudes and impressions changed for planetarium visitors minute-by-minute. Attendee satisfaction was becoming a high priority because Africk (1971) estimated that there were more than 1500 functioning planetarium facilities in the United States with a total estimated annual attendance at over 20,000,000. Prior to Gutsch's (1978) innovative work, planetarium education researchers relied most heavily on exit surveys to understand the attendees' experiences. This snapshot exit strategy for planetarium education research has obvious disadvantages because it is impossible to dissect subcomponents of the larger 40-min long planetarium experience. However, during this time, the academic disciplines of mass-media, marketing and political science research were beginning to learn more about their audience by using electronic voting devices. Called portable, electronic feedback systems (Fig. 5.2), these push button

Fig. 5.2 Personal attitude response device with *Red* (on *left*) and *Green* (on *right*) buttons constructed by Gutsch (1978)

devices were attached to the then newly available computers and could create second-to-second recordings of responses throughout the entire duration of a program.

This system designed and assembled by Gutsch (1978) was able to record the attitudes of planetarium visitors throughout a planetarium program. Visitors would hold the device in their lap and press a green button when they felt a positive attitude for something they liked (perhaps a special effect showing a shooting star) and, alternatively, press the red button when they felt a negative attitude for something they disliked (perhaps the volume became uncomfortably loud). Audience members were told to leave the buttons un-pushed during periods of the show that did not evoke strong positive or negative emotions. In essence, audience members were able to report: (i) like; (ii) neutral; and (iii) dislike.

All of the green (positive) and red (negative) button pushes were recorded by the computer and given a specific time stamp. Then, he was able to carefully correlate each individual element of the show occurring at a particular instant in time with up to 20 attendees' positive or negative impressions of that specific element. The result was a detailed graph of changing audience attitudes throughout the show (an example is shown in Fig. 5.3). The produced chart is similar to the contemporary, more familiar like-and-dislike time charts that are produced for political analysts by audience members listening to a spirited debate between candidates competing for political office.

Armed with this sophisticated measurement tool, Gutsch (1978) was able to make highly unique measurements of visitor attitudes about audio and visual stimuli in planetarium programs at the 65-foot diameter, 240-seat Strasenburgh

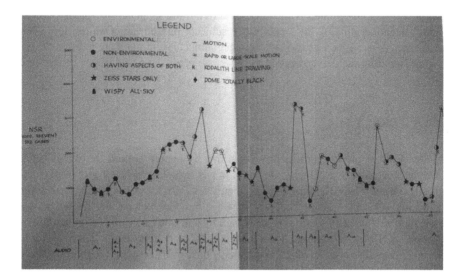

Fig. 5.3 Real-time chart strip marked for planetarium show element and participants' attitudes (reprinted with permission from Gutsch 1978)

Planetarium in Rochester, New York. Gutsch (1978) was able to conduct many experimental trials looking at different variables relatively quickly. He looked at the responses of 592 voluntary study participants over 41 trials, yielding an enormous dataset.

The data showed that participants responded most positively to whole star-filled dark skies with the Milky Way visible and special effects visuals with motion (spinning black holes, fluctuating aurora, flying rockets, and shooting stars). In contrast, participants most negatively viewed static or colorless visuals and static visuals that were projected for extended amounts of time, as well as heavily accented narration (e.g., Herschel). In the typical planetarium program, some visual effects are used repeatedly throughout the show. Participants would often identify a particular visual as positive the first time it was used, but then not the second or third time it was used, perhaps because they were habituating to the effect or because they thought they had already recorded their pleasure (or displeasure) with the event. Participants often reacted negatively to loud audio effects, such as those that accompany descriptions of the creation of the universe or the last death throes of a supernova.

Some "events" defy a singular classification as only music, sound effect, or visual, because they are extended and combined effects. An example of this is a sequence depicting early determinations of the speed of light, or a sequence depicting the *Titus-Bode rule* identifying the relative distance of planet from the Sun where numbers appear on the dome much like an animated chalk board. Both of these extended events, when used in these experiments, were correlated with negative attitudes by most attendees, suggesting that math should be used sparingly in planetarium shows.

Perhaps surprisingly, Gutsch (1978) also found that age as a variable relates to the overall response patterns in terms of activity and tendency for participants to record negative attitudes. For one, older subjects pushed either button less often than younger subjects. For another, younger subjects were more apt to press the red button signaling dislike, compared to older subjects.

High Value Visits

At the peak of planetarium facility construction, people around the world began studying the impact a planetarium visit, or visits, could have on individuals and groups. Firniss (1980), noting that the there was little to no published planetarium education research outside the US, despite that the planetarium projector was essentially invented in Germany, surveyed 70 individuals visiting planetariums in Kuwait, Stuttgart, Nurnberg, Vienna, Bangkok, and Reutlingen. There were 13 affective domain items in which respondents universally reported high educational value for their planetarium learning experiences, no matter what country they were in when surveyed. As a second sub-study, he surveyed 222 high school students in West Germany on a 10-item, Likert-style survey who also almost universally responded positively to the notion that the planetarium is an enjoyable and valuable learning experience.

Mallon's 1980 Survey

After 15 years of affective domain research in planetarium education, researchers were finding that people who visited a planetarium generally enjoyed their experience, but the degree to which life-altering attitudes were being impacted was mixed, at best. To gain a better understanding of the landscape of planetarium education for impacting attitudes, values and interests of students visiting planetariums, Mallon (1980), who was a doctoral student at Temple, devised a comparative study of five locations distributed across the US. Published by Mallon and Bruce (1982), the study involved 556 8–10 year old students: 324 from Methacton, PA; 76 from Richardson, TX; 52 from Berkeley, CA; 52 from St. Paul, MN; and 52 from Reno, NV.

Mallon (1980) adapted an attitude survey that had been developed earlier by Fisher (1973) that contained 20 Likert-style items with statements respondents were to judge the extent to which they agreed with (Table 5.3). The survey itself was not distributed to students, but instead read aloud in an effort to mitigate against the confounding variable of unequal reading levels within the groups.

To conduct the comparison study, two common 40-min program scripts that used both the star projector and a 35 mm slide projector were developed that focused on the same astronomical concepts: one was focused on being an interactive, participatory program whereas the other program was a traditional, information-download planetarium lecture. Although the authors come to the conclusion that the planetarium is possibly effective and has enhancing affective characteristics, Mallon's (1980) data only weakly confirms this as, with only few exceptions, students who attended either the traditional planetarium lecture or the participatory planetarium program appeared to have about the same middle range attitude scores before and after their learning experiences. As further evidence, Edoff (1982) found the same result of no changes in attitudes observed insofar as the instrument could measure when he used Mallon's (1980) attitude survey, which he curiously called an *opinionnaire*, as part of a two-group, pretest posttest comparison study, to comparatively detect changes in attitudes when students actively used manipulatives in the planetarium versus students who did not use manipulatives.

After the 1980s' Hiatus

As with planetarium education research studies focusing on the cognitive domain of memory and understanding, there was very little published planetarium education research on the affective domain of attitudes, values, and interests during much of

Table 5.3 Likert-style affective domain survey adapted from Mallon (1980)

Statement	Strongly disagree	Disagree	Undecided	Agree	Strongly agree
1. Reading about Astronomy is difficult	SD	D	U	A	SA
2. We spend too much time doing experiments	SD	D	U	A	SA
3. I am learning a lot about Astronomy this year	SD	D	U	A	SA
4. What we do in the planetarium is what a real astronomer would do	SD	D	U	A	SA
5. In the planetarium, we study 'Today's Problems'	SD	D	U	A	SA
6. I dislike coming to the planetarium	SD	D	U	A	SA
7. I read more astronomy materials than I did last year	SD	D	U	A	SA
8. I enjoy doing the astronomy	SD	D	U	A	SA
9. I can solve problems better than before	SD	D	U	A	SA
10. My friends enjoy doing the astronomy experiments	SD	D	U	A	SA
11. What I am learning in the planetarium will be useful to me outside school	SD	D	U	A	SA
12. I think about things we learn in the planetarium when I'm not in school	SD	D	U	A	SA
13. I do not want to take any more astronomy classes than I have to take	SD	D	U	A	SA
14. Reading astronomy is more fun than it used to be	SD	D	U	A	SA
15. Experiments are hard to understand	SD	D	U	A	SA
16. Astronomy is dull for most people	SD	D	U	A	SA
17. The things we do in the planetarium are useless	SD	D	U	A	SA
18. The kinds of experiments I do in the planetarium are important	SD	D	U	A	SA
19. I learn a lot from doing my astronomy experiments	SD	D	U	A	SA
20. Most people like planetarium lessons	SD	D	U	A	SA
	Strongly disagree	Disagree	Undecided	Agree	Strongly agree

the 1980s decade. When published research reappeared at the end of that decade, there was a technologically brand new planetarium world available, as technology had made tremendous leaps in how planetariums projected visuals on the dome. There were also dramatic differences in terms of moving from celestial sphere dominated programs to current-research results in astronomy, planetary geology, and space science. The change in what was being taught in planetariums also impacted the types of planetarium education research questions being pursued and the research methods which we used by planetarium education researchers.

The Portable Theater

One of the technological advances occurring at the end of the Century was a dramatic decrease in construction of brick-and-mortar planetarium facilities and an increase in the availability of lower-cost, portable planetariums. A portable plan-etarium was not only available at a lower cost, it could be transported from one school building to the next, lowering overall school bus transportation costs. Moreover, portable planetarium programs were largely self-contained and could be operated—at highly varying levels of quality—by minimally trained elementary school teachers, thereby reducing schooling costs considerably by not having to hire a full-time planetarium educator.

Initially, the most popular of the portable planetarium systems was *STARLAB*. Taking advantage of the STARLAB portable planetarium system, and a decade after Twiest (1989) found that the plantarium struggled to better enhance student attitudes as compared to classroom instruction, Meyer (2000) used Twiest's (1989) survey instruments to study the value of pre- and post-planetarium visit learning activities when using inflatable, portable *STARLAB* planetarium. She designed a multiple treatment, pre-test posttest quasi-experimental study design to see if 6th grade students who participated in various pre- and post-visit activities would have enhanced attitudes toward astronomy. She used a 13-item, pretest-posttest, Likert-style survey (shown in Table 5.4) to determine changes in students' attitudes based on three different treatment conditions. One treatment group of 56 students experienced hands-on activities before and after their planetarium visit, a second treatment group of 60 student used audio-visual activities before and after their visit, and a third treatment group of 65 students completed text and reading based activities before and after the planetarium. ANOVA and Mann-Whitney U statis-tical test analysis suggested that there were no significant differences among any of various treatment groups. These results are consistent with what Wright (1968) found decades earlier: that although pre- and post- field trip activities do appear to be important for many field trip learning experiences, it had not been shown to be true for visits to the planetarium.

In addition to the Likert-style attitude survey toward astronomy shown in Table 5.4, Meyer (2000) also wanted to get a sense of which instructional approaches students most enjoyed and found to be most valuable. She describes

Table 5.4 *Astronomy attitude inventory* used by tweist (1989) and Meyer (2000) Adapted from Moore and Sutman (1970)

Statement	Strongly disagree	Disagree	Undecided	Agree	Strongly agree
Practice 1: I like candy	*SD*	*D*	*U*	*A*	*SA*
Practice 2: I don't like dogs	*SD*	*D*	*U*	*A*	*SA*
1. I like studying science	SD	D	U	A	SA
2. The planetarium is an interesting place to me	SD	D	U	A	SA
3. I would like to go to the moon	SD	D	U	A	SA
4. Scientists always want to learn new things	SD	D	U	A	SA
5. Science is so hard that only very smart people can do it	SD	D	U	A	SA
6. Most people are able to learn science	SD	D	U	A	SA
7. I don't' like astronomy because I think space is boring	SD	D	U	A	SA
8. Science is too hard for me	SD	D	U	A	SA
9. Learning about space is exciting to me	SD	D	U	A	SA
10. The things I learn in the planetarium are fun	SD	D	U	A	SA
11. I like to go outside and look at the stars	SD	D	U	A	SA
12. Scientists do not have enough time for fun	SD	D	U	A	SA
13. Of all my classes, science class is one of the classes I like the most	SD	D	U	A	SA
	Strongly disagree	Disagree	Undecided	Agree	Strongly agree

piloting a 25-item survey on "How I Feel About Today's Lesson" adapted from earlier work but found it far too cumbersome to use with elementary grades students. Instead, she asked students three questions to gain insight into students' perceptions of the different instructional approaches used in the study. The questions were:

[1] *The thing I like most about today's lesson is* _____.
[2] *If anything, I did not like* _____.
[3] *Would you want to do this again?* [circle] YES NO

As with her affective domain survey (and a cognitive domain survey as well), she was not able to discern any differences among student responses regardless of which educational intervention her 6th grade students participated.

One might question why Meyer (2000) and other highly competent and innovative educators of the day seemed to be detecting any observable differences in learning experiences. A closer look at the attitude survey instrument Twiest (1989) and Meyer (2000) used provides shows that using surveys such as these were a fairly common practice and form among science education researchers in the 1970s and early 1980s. Their survey used in this study was a modified version of a survey created by Moore and Sutman (1970) decades earlier. Twiest (1989) modified it by changing the wording of particular phrases to focus on astronomy and the planetarium specifically instead of science, in general. Additionally, because Moore and Sutman (1970) validated the survey for use with high school students who have a more extensive and nuanced reading ability, Meyer (2000) further modified it by using more straightforward and simple language so that it could be used by the elementary-level students in her study. The overarching problem is that a scholar does not know how changing the specific words in a previously used survey changes the reliability and validity of the instrument, and such changes require the instrument to be re-validated before use.

As a further problem, the widely used science attitude survey published by Moore and Sutman (1970) was found to be highly inconsistent, and criticized strongly by scholars such as Munby (1983). This criticism led to a drastic reworking of the original science attitude survey by Moore and Fox (1997) twenty years after its original insertion into the science education research community. However, because Meyer (2000) was relying heavily on Twiest's much earlier work (1989), she may have ended up using an old, out-of-date survey that had long since been replaced with a newer survey by the contemporary research community, of which she was likely unaware.

Reassessing the Research Instruments

The largely disappointing planetarium education research results consistently seen up to this point in history highlights the overarching difficulty faced by planetarium education researchers when it comes to measuring variables in the affective domain, as well as the cognitive domain: The measurement instruments most widely used by planetarium education researchers seemed to lack sufficient resolution power to observe any of the cognitive or affective changes planetarium education advocates were convinced were there. A complete deficit of appropriate educational survey instruments widely agreed upon across the broad planetarium education research community has greatly hampered planetarium education research, resulting in most studies showing no statistically significant results observed, insofar as the available instruments could measure. Planetarium education researchers tried to mitigate for the lack of planetarium-specific instruments by trying to adapt existing instruments used by science education classroom researchers. However, many of these instruments too often themselves lacked strong foundations in reliability and validity. Moreover, even when the best of those instruments meet the basic criteria of fair assessment measures, they were usually validated for science classroom environments, making them suspect for making precise measurements of student achievement and attitudes when used in a seemingly very different planetarium teaching environment. Furthermore, when planetarium education researchers were unsuccessful at modifying existing instruments constructed for purposes different than the planetarium learning environment, they resorted to creating their own survey instruments. Perhaps unbeknownst to the well-intentioned scholar, the development and validation of reliable measurement instruments is an incredibly arduous and time consuming task if it is to meet both the criteria for good assessment and have the needed resolution power to see subtle changes in students' none-too-obvious mental landscapes. The end result of all of this is that considerable time and effort was likely largely misplaced trying to make subtle measurements using insufficient instrumentation resulting in the large number scholars repeatedly reporting "no statistically significant differences observed."

New Directions

By the turn of the century, planetarium education researchers started using a larger research toolkit for studying the seemingly difficult to precisely quantify affective domain. The study methods from the qualitative and interpretive research community provided new and desperately needed pathways to gain deeper insight into how planetarium visitors make meaning of their experiences when learning in the planetarium (viz., Plummer et al. 2015). As an example, while working on her Master of Arts in Reading Education focusing on storytelling, Meyers (2005) began exploring how pairing folk tales and scientific explanations might capture students' interests and enhance their attitudes. Coming from the reading education research tradition, as opposed to the science education research tradition that provides the scholarly foundation for most planetarium education researchers, she was able to study the impact of storytelling and scientific explanations on the attitudes of 3500 3rd–6th elementary students at eight locations using a portable planetarium with multiple measures. The overarching research question guiding her study was, "Will there be a difference in student attitudes before and after experiencing a *SkyTeller* [paired story and scientific explanation] module?" She created a 10-item, Likert-style, pretest-posttest survey to assess students' attitudes toward science and stories as had been done in many earlier planetarium education research projects; however, she also triangulated her results using questionnaires, telephone interviews, and field-notes made during on-site visits with the various program presenters to gain a deeper understanding of the 3rd–6th grade student experience, as well as boost confidence in her conclusions.

Funded in part by the US National Science Foundation, students studying space science as part of this study participated in 10 developed program modules, each lasting 12–15 min. Each module featured a Native American story about a topic that is included in the domain of astronomy, followed by a science narrative using imagery from the Native American story. The underlying educational innovation notion is to present the science of astronomy in the framework of a story inside the planetarium instead of the more typical "tightly worded fact and inference mode of the science essay" lecture (Meyers 2005, p. 9).

Changes in student attitudes were initially measured using a pretest and posttest survey. She piloted multiple versions of the survey, finding that vague statements such as "I don't like science" lacked the measurement sensitivity as emotionally vectored statements such as "Science is too hard for me." Although visual inspection can initially show positive changes in student scores, statistically significant differences are only sometimes mathematically confirmed. Part of the challenge Meyers (2005) encountered early in her study was that she had a stack of student pre-tests and a stack of student posttests that were completely unmatched. Because she did could not tell specifically how much each individual student responses changed from pretest to posttest, she had to use independent samples t-tests for significance.

She also used MANOVA tests of variance to try to account for unknown confounding variables at the different study sites that could have impacted results. If she had known each student's individual changes, by matching each student's pretest to their specific posttest by including their name or a confidential identifier, she could have used a paired-samples t-test of significance, which is not nearly as stringent in its mathematical requirements to eliminate the role of chance when comparing groups. The advantage to not putting identifiers on each test is that it is often easier to acquire institutional review board approval for research on human subjects when dealing with minor children and no demographic or identifying information is collected. The disadvantage is that less powerful statistical measures are the only thing that is available to the researcher. Prior to 2000, institutional review board (IRB) approval for research on human subjects was only occasionally required for similar education research projects, but after 2000 such institutional reviews became more commonplace.

By the end of the project, Meyers (2005) was able to obtain large enough sample sizes (N = 884) to confirm that students' attitudes toward science and toward story increased in amounts that were quite small in effect size, but statistically significant. These results were buoyed upward by supporting telephone interviews, individual interviews, and excerpts from field-notes she collected. She concluded that couching science in a wrapper of storytelling was an important innovation in planetarium education, and is consistent with arguments advanced by other scholars (Slater 2014a, b). Perhaps more important than her results, was a change in how research could be done by planetarium education researchers. In stark contrast to many of the earlier studies in planetarium education research, Meyers (2005) studied the impact not of her own teaching, but that of others. This is evidence of a larger paradigm shift from the dominant practice of researching one's own teaching

or one's own planetarium programs to instead researching the educational trans-
formation of students in presentations other than ones given primarily by the
researchers themselves.

However, not all planetarium education researchers were adopting the new
methods of qualitative and interpretive research. Platco (2005) conducted a
two-group comparison study of 295 8th grade students in Pennsylvania who
attended either a traditional "star show" planetarium lecture or a "participatory
oriented planetarium" program. His study was similar to a decades earlier research
study by Bishop (1980), which aimed to determine which characteristics made a
planetarium show more effective than other planetarium shows. The study was
ultimately unable to detect differences in achievement insofar as the instruments
could measure. Somewhat suspect as to its appropriateness, he used a pre-existing
22-item attitude survey on his 8th grade student participants that was designed
specifically for college students by Zeilik (1999) and colleagues. He found that both
treatment groups had small, but statistically equivalent increases, regardless of
which type of show they experienced, changing from a mean of 3.2–3.4 on a
5-point, Likert-style scale.

South African Mixed-Methods Study

As part of a much larger study in South Africa, Lelliot (2007) leveraged a
mixed-methods study approach to pursue a better understanding of students'
evolving attitudes about astronomy. He evaluated surveys, coded statements from
interviews, and diagnosed graphical concept maps drawn by his 7th, 8th, and 9th
grade student-participants who visited a number of astronomy-related scientific and
public information sites, including the Johannesburg Planetarium, located on the
campus of the University of the Witwatersrand. The planetarium program offered a
show of about one hour in which the planetarium educator demonstrates the solar
system as seen from Earth via a video projection on the dome (the planetarium was
at the time a classical opto-mechanical, non-digital system) and from outer space,
the nature of astronomical distances, and the night sky constellations visible in the
current night's sky (Table 5.5).

The combination of interviews, surveys, and graphic concept maps provided
significantly more results about students' changes in the cognitive domain than the
affective domain. Nonetheless, the sum total of data collected by Lelliot (2007)
showed that students enjoyed the planetarium and desired to visit again, with their
families or fellow school mates. The most important aspects of the planetarium
presentation itself that most positively enhanced student's attitudes across the
dataset were observations of the stars in the night sky, the notion that stars spin,
names and dot-to-dot outlines of constellations, and a special effects, scientific
visualization of the formation of planetary systems from enormous dust clouds.

Table 5.5 Lelliot's (2007) abridged interview protocol

I. Pre- and Post-planetarium visit knowledge questions

1. Do you know the names of any stars? Can you see any planets in the night sky? What do they look like?
2. What is the solar system? What ['things'] does it consist of? What shape is it? How do you know?
3. Tell me anything you know about the sun
4. Why does the sun move across the sky every day?
5. The sun and the moon look the same size in our sky. Are they? Why do you say that? [Which is really bigger?] Why do they look the same size?
6. Stars at night look like pinpricks of light. Why? What are they? [Why are they there?]
7. In the fastest space ship we could build, how long would it take to reach the closest star outside our solar system? Guess?
8. Why does the moon look different on different nights (and days) throughout the month?
9. Have you heard any planet or space-related things in the news at all recently?

II. Pre-Planetarium visit attitude interview questions about science and astronomy

1. Do you like school? Probe why/not
2. What are your 2 favorite subjects at school? Why?
3. What are your 2 least favorite subjects at school? Why?
4. What do you do in your spare time? What TV programs do you like? Do you read much? Tell me any books you have read recently
5. So far in your schooling, have you ever covered things like the sun, the moon, planets, space, stars etc.? In which subject did you cover them?
6. Do you find anything to do with space, planets and stars interesting? If so, what? Why? If not, why not?
7. If you were a teacher teaching about space, planets and stars, what would you do with your students (in your own grade)?
8. Do you believe in aliens? Probe why. Have aliens ever visited earth? Have scientists found life anywhere other than earth?
9. Do you believe in the horoscope in magazines etc.? [Astrology]. Probe
10. Are you religious? What is god's relationship to the universe?

III. Post-Planetarium visit attitude interview questions about science and astronomy

1. Now you've been to the planetarium, have your ideas about space, planets and stars changed at all? What?
2. What things did you most enjoy about the visit? Can be anything
3. What did you dislike about the visit?
4. Have you told anyone about the visit? Who? What did you tell them?
5. If your school arranges a visit to Planetarium, would want to you go (again)?
6. The Planetarium runs public visits. Would you want to visit again together with your family?
7. What sort of job do you want to do when you leave school?
8. Other than today, have you thought about the trip since? Tell me what

Digital Theaters and New Research

At the end of the 21st Century's first decade, Bruno (2008) estimated that there were more than 3000 planetarium facilities worldwide, and that nearly 600 of those had computer-driven, full-dome video projection systems. Despite the unlimited number of options of what specific scientific notions can be projected on the dome, there was still was great interest among planetarium educators specifically to use the planetarium to enhance the affective domain of visitors. Small and Plummer (2010) specifically targeted figuring out what the broader community of planetarium educators' goals were for their audiences. Twenty-five male and 11 female planetarium directors, operators, and vendors were interviewed, more than half of which had been in the planetarium field for longer than 15 years. The resulting interview transcripts were inductively coded using a common practice, constant comparison method in the spirit of Strauss and Corbin (2014). The two dominant themes across all interviewees were (i) increasing knowledge about specific astronomical concepts and (ii) increasing interest in and enhanced awareness of astronomy.

These results add weight to the notion that planetarium educators do not just want visitors to leave knowing more astronomy: They also want visitors to like astronomy, value astronomy, and be interested in astronomy as a result of engaging with the planetarium programming. Plummer and Small (2013) extended their interview analysis of planetarium educators' perspectives on increasing the audiences' interests in astronomy. Three emerging ideas came out of interview analysis. The first is that planetarium educators have a broad-based desire to enhance the levels of enthusiasm or interest of their audiences, although they are often unable to articulate a clear motivation for why such would be important. The second emerging idea is that planetarium educators hold a great desire to increase interest and engagement for the reason that it can better foster learning. The third emerging notion from the data was that planetarium educators have a desire to increase interest among their visitors so that the planetarium experience can serve as an impetus to extent and pursue learning opportunities beyond any one specific planetarium experience. Taken together, Plummer and Small's (2013) data shows that there is a consistent believe across the planetarium education community that an effort to enhance visitor's affect has the potential for long-term education impacts and benefits, albeit largely intangible.

Schmoll (2013), an astronomy graduate student at the University of Michigan, wanted to explore a number of cognitive and affective aspects of learning in planetariums, and created an extensive and complex set of curriculum interventions based on a solid theoretical foundation. For the affective component to her work, she conducted a single-group, posttest only study of 5th grade students from a public elementary school in southeast Michigan who visited a digital planetarium at a local natural history museum. The 29 5th grade students in her study population had similar demographics to the school district's overall socioeconomically, affluent population: 95% Caucasian, 2% Asian/Pacific Islander, 1% Black, 1% American Indian/Alaskan Native, and 1% Hispanic. Students were given a 15-day instructional unit on astronomy that included pre- and post-planetarium activities in which the planetarium field trip occurred near the middle of the unit.

The Likert-style survey Schmoll (2013) adopted had four statements for students to judge the extent to which they agreed or disagreed: (1) *I enjoyed the planetarium visit* [measuring attitude about the planetarium experience]; (2) *The planetarium was interesting* [measuring interest in the planetarium]; (3) *I would like to learn more about astronomy* [measuring value in astronomy]; and (4) *I think astronomy is*

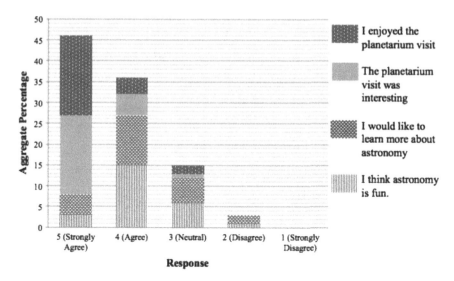

Fig. 5.4 Students' affect after visiting a planetarium [adapted from Schmoll (2013, p. 96)]

fun [measuring attitude about astronomy]. Illustrated in Fig. 5.4, the survey results revealed that 46% of students strongly agreed with all provided statements, and 82% either agreed or agreed strongly with the statements. Viewing these survey questions and results through a critical lens causes one to wonder if these same results would have occurred for nearly any out of classroom learning experience that is different than the day-in and day-out schooling experience such as visiting a forensic crime lab, an archeological dig site, or a state wildlife reserve. In other words, a more useful measure might be to build less vague questions that provide at least some specific insights about the planetarium experience that would point a planetarium education researcher toward things to keep about the planetarium learning experience and what to alter.

That same year, Petrie (2013), a graduate student in museology at the University of Washington, sought to better understand why families with pre-school students attended planetariums and what parents thought their pre-school aged children were learning by attending a planetarium show. She conducted short pencil-and-paper surveys and short interviews with children ages 3 to 5 years and their parents in family groups after they watched an early-childhood oriented planetarium show. The planetarium show emphasized the following three concepts: (i) the Moon is visible during the day; (ii) the sky contains the Sun, Moon, and stars, which many people understand by telling stories; and (iii) astronauts move differently because there is less gravity on the Moon. She collected 57 questionnaires and conducted 12 interviews.

Summarized in Table 5.6, the survey results showed that the most common reasons parents were motivated to attend the early-childhood planetarium show were because they wanted their child to learn about astronomy and they themselves

Table 5.6 Reasons sparents attend an early-childhood planetarium show with their children (Petrie 2013, p. 25)

Reason	Percentage of surveys including this response (%)
I wanted my child to learn about astronomy	80
I was personally interested in the program and/or astronomy	75
My child is interested in astronomy	49
This show was happening at a convenient time	47
The show is free [of charge]	37
Other (Write-in) [unclassified response]	11
I wanted to rest while my child does an activity	9
I wanted my child to feel more comfortable in the dark	5

were personally interested in astronomy. Other commonly described reasons were that the show was happening at a convenient time and that parents needed an opportunity to rest while their child did an activity that did not require their supervision or participation. These results do not provide insight into how attitudes of the participants changed as a result of the program; however, they provide tremendous insight into what motivates parents and possibly can be leveraged by planetarium educators to increase their attendance. In this sense, the demarcation between the scholarly fields of marketing and affective-focused planetarium education research are somewhat blurred.

Conclusion

In the early days of planetarium education research, most studies involved figuring out how much planetarium attendees remembered about the names and positions of constellations and what they now better understood about the predictable movements on the celestial sphere and the planets. These are all aspects of the cognitive domain of learning. As early studies showed little dramatic changes in attendees' cognitive domains as a result of being inside the planetarium, planetarium education researchers began to wonder if the strongest and most measureable changes resulting from learning in the planetarium occurred in the affective domain of enhanced attitudes, values, and interests. This idea of expanding the populations' value of all-things-space would be consistent with the goals of the US 1958 National Defense Education Act that contributed significant funding and drove the construction of first hundreds then thousands of planetarium facilities; however, this required planetarium education researchers to adopt new research methods from the qualitative and interpretive research paradigms (Fig. 5.5). What we now have come to better appreciate is that the planetarium in and of itself in isolation is not a

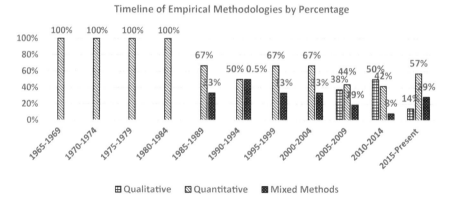

Fig. 5.5 Evolving use of different empirical research methods

magical silver-bullet for solving all of astronomy education's challenges for improving learning and attitudes. Instead, planetarium education programs need to use the same educational theory-driven, research-confirmed best practices in science education to help enhance learners' cognition and affect. The planetarium is unarguably able to capture attendees' innate interest, but planetarium education research confirms that lasting change requires purposeful educational decisions in order to be relevant and effective.

The presentation of planetarium shows has undergone a paradigm shift in the early 21st century with its move to digital projection systems and use of actual data embedded in the real-time software, opening avenues for interesting future research. The ability for spatially powerful demonstrations opens the doors to inspired educational research in the effectiveness of learning for those students whose spatial abilities are strong. In addition, the impact of using actual data in live planetarium shows as opposed to artistically rendered scenes is another area ripe for new research. This research is needed to help drive the best practices in using the powerful spatial domain of the digital planetarium revolution.

References

Africk, S. A. (1971). Where Is astronomical education headed? *Sky and Telescope, 42,* 277.
Anderson, L. W., & Bourke, S. F. (2000). *Assessing affective characteristics in the schools.* Routledge Publishing.
Bishop, J. E. (1980). *The development and testing of a participatory planetarium unit employing projective astronomy concepts and utilizing the karplus learning cycle, student model manipulation and student drawing with eighth-grade students.* Doctoral Dissertation, University of Akron.
Bruno, M. (2008). *Trends in full-dome show production and distribution.* Paper presented at the International Planetarium Society Meeting on July 3, 2008 in Chicago, IL.

Burnette, W. N. J. (1976). *Use of the planetarium in changing attitudes and achievement in earth-space science education.* Doctoral Dissertation. University of Missouri, Columbia.

Corbin, J., & Strauss, A. (2014). *Basics of qualitative research: Techniques and procedures for developing grounded theory.* Sage Publications.

Edoff, J. D. (1982). *An experimental study of the effectiveness of manipulative use in planetarium astronomy lessons for fifth and eighth grade Students.* Ed.D. Dissertation. Wayne State University.

Firniss, K. P. (1980). *The application of the planetarium as an education tool.* Doctoral Dissertation. Open University, Institute of Educational Technology, England.

Fisher, T. H. (1973). The development of an attitude survey for junior high science. *School Science and Mathematics, 73*(8), 647–652.

Gable, R. K., & Wolf, M. B. (2012). *Instrument development in the affective domain: measuring attitudes and values in corporate and school Settings* (Vol. 36). Springer Science & Business Media Publishing.

Gardner, M. H. (1964). The planetarium as an educational tool. *The Science Teacher, 31,* 17–18.

Gutsch, W. A. (1978). *Obtaining and analyzing affective response profiles in a planetarium environment: An exploratory study.* Ph.D. Dissertation. University of Virginia.

Hurd, D. W. (1997). Novelty and it's relation to field trips. *Education, 118*(1), 29.

Jamison, M. M. (1972). *A consideration of the planetarium and the lecturer as agents to effect change in administrators regarding social attitudes in school and community.* Ph.D. Dissertation. University of Illinois, Urbana-Champaign.

Lelliott, A. D. (2007). *learning about astronomy: a case study exploring how grade 7 and 8 students experience sites of informal learning in South Africa.* Doctoral dissertation. University of the Witwatersrand, Johannesburg.

Mallon, G. L. (1980). *Student achievement and attitudes in astronomy: an experimental study of the effectiveness of a traditional "star show" planetarium program and a "participatory oriented planetarium" program.* Ed.D. Dissertation. Temple University.

Mallon, G. L., & Bruce, M. H. (1982). Student achievement and attitudes in astronomy: An experimental comparison of two planetarium programs. *Journal of Research in Science Teaching, 19*(1), 53–61.

Meyer, J. R. (2000). *The effects of three types of pre-and post-planetarium/STARLAB-Visit instruction methods on astronomy concepts and attitudes of sixth grade students.* Ph.D. Dissertation. University of Missouri—Kansas City.

Meyers, M. B. (2005). *Telling the Stars: A Quantitative Approach to Assessing the Use of Folk Tales in Science Education.* Master of Arts in Reading Education Thesis, East Tennessee State University.

Moore, M. G. (1965). *An analysis and evaluation of planetarium programming as it relates to the scienceeducation of adults in the community.* Ph.D. Dissertation, Michigan State University.

Moore, R. W., & Sutman, F. X. (1970). The development, field test and validation of an inventory of scientific attitudes. *Journal of Research in Science Teaching, 7*(2), 85–94.

Moore, R. W., & Foy, R. L. H. (1997). The scientific attitude inventory: A revision (SAI II). *Journal of Research in Science Teaching, 34*(4), 327–336.

Munby, H. (1983). Thirty studies involving the "Scientific Attitude Inventory": What confidence can we have in this instrument? *Journal of Research in Science Teaching, 20*(2), 141–162.

Petrie, K. B. (2013). *Early childhood learning in preschool planetarium programs.* Master's Thesis. University of Washington.

Platco, N. L. (2005). *A comparative study of the effectiveness of "star show" vs." participatory oriented planetarium" lessons in a middle school STARLAB setting.* Ed.D. Dissertation. Temple University.

Plummer, J. D., Schmoll, S., Yu, K. C., & Ghent, C. (2015). A guide to conducting educational research in the planetarium. *Planetarian 44*(2), 8–24, 30.

Plummer, J. D., & Small, K. J. (2013). Informal science educators' pedagogical choices and goals for learners: The case of planetarium professionals. *Astronomy Education Review, 12*(1), 010105.

Reed, G. (1970). *A comparison of the effectiveness of the planetarium and the classroom chalkboard and celestial globe in the teaching of specific astronomical concepts.* Doctoral Dissertation. University of Pennsylvania.

Reed, G. (1972). An outline for the education of a K-12 planetarium teacher. *Journal of College ScienceTeaching, 1*(3), 51–52.

Reed, G. (1973). The planetarium versus the classroom—An inquiry into earlier implications. *School Science and Mathematics, 73*(7), 553–555.

Reed, G. (1975). The affective value of a planetarium in the scheduling of a college astronomy course. *School Science and Mathematics, 75*(8), 716–722.

Reed, G., & Campbell, J. R. (1972). A comparison of the effectiveness of the planetarium and the classroom chalkboard and celestial globe in the teaching of specific astronomical concepts. *School Science and Mathematics, 72*(5), 368–374.

Ridky, R. W. (1974). *A study of planetarium effectiveness on student achievement, perceptions and retention.* Paper presented at the National Association for Research in Science Teaching, Chicago, IL. Available online at: http://eric.ed.gov/?id=ED091207

Schmoll, S. E. (2013). *Toward a framework for integrating planetarium and classroom learning.* Ph.D. Dissertation. University of Michigan.

Sinatra, G. M., Broughton, S. H., & Lombardi, D. (2014). Emotions in science education. *International Handbook of Emotions in Education* 415–436.

Slater, T. F. (2014a). *How much louder do I need to turn up the soundtrack before they learn?: A cognitive science perspective on memory in the planetarium.* Proceedings of the Great Lakes Planetarium Association, Muncie, IN, October 30, 2014, retrieved from YouTube.com at: https://www.youtube.com/watch?v=Vu9MzFIwJMM

Slater, T. F. (2014b). *How to make ASTRO 101 classes more memorable.* Blog post retrieved from astrolrner.org at: https://astronomyfacultylounge.wordpress.com/2014/12/01/how-to-make-astro-101-classes-more-memorable/

Slater, S. J., Tatge, C. B., Bretones, P. S., Slater, T. F., Schleigh, S. P., McKinnon, D., et al. (2016). iSTAR First light: Characterizing astronomy education research dissertations in the iSTAR database. *Journal of Astronomy and Earth Sciences Education, 3*(2), 125–140.

Small, K. J., & Plummer, J. D. (2010). Survey of the goals and beliefs of planetarium professionals regarding program design. *Astronomy Education Review, 9*(1), 010112.

Smith, P., & Ragan, T. J. (1999). *Instructional design.* New York: John Wiley & Sons.

Twiest, M. G. (1989). *The attitudinal and cognitive effects of planetarium integration in teaching selected astronomical concepts to fourth, fifth, and sixth grade students.* Doctoral Dissertation. University of Georgia.

Vance, J. B. (1964). A community uses the planetarium. *The Science Teacher, 31*(6), 19–20.

Wright, D. L. C. (1968). *Effectiveness of the planetarium and different methods of its utilization in teaching astronomy.* Doctoral Dissertation. University of Nebraska.

Yee, A. H., Baer, J. M., & Holt, K. D. (1968). *An evaluation of the effectiveness of school planetarium experiences.* Educational Technology Publications, as cited by Firniss, K. P. (1980). *The Application of the Planetarium as an Education Tool.* Doctoral Dissertation, Open University, Institute of Educational Technology, England.

Zeilik, M., Schau, C., & Mattern, N. (1999). Conceptual astronomy. II. Replicating conceptual gains, probing attitude changes across three semesters. *American Journal of Physics, 67*(10), 923–927.

Chapter 6
Epilogue

Since the 1930s, the detailed and scholarly efforts to perform education research in the classical planetarium field have been remarkable. However, it is worth pointing out that an Earth-centered view of the universe offered by optical star projectors provides a view of the universe that is inherently flat to the student. Thus, extra effort is required to convey the true three-dimensional nature of the cosmos and the complicated spatial relationships that exist therein.

From the introduction of digital planetariums in the late 1990s, planetariums' usage of the digital domain has matured to a level where most can now convey what Edwin Hubble discovered in the 1930s (ironically just at the time the optical star projector was introduced), that island universes are truly galaxies, and exhibit real astronomical data directly onto the dome, from surfaces of planets and moons, out to very great distances with surveys like the Sloan Digital Sky Survey. This advancement in technology has allowed a representation, in a virtual simulation using real data, of the distances between galaxies and the Milky Way's place in the cosmos for the very first time in human history.

This new view of astronomy, as presented on planetarium domes, displays cosmological content as well as planetary and stellar science. Planetariums provide a golden opportunity for some new efforts in educational research and encourage the development of new tools that offer a deeper understanding of the cosmos by students and the public.

It is both necessary and important to continue researching scholarly educational in the planetarium setting. New methods and techniques are needed in order to properly teach the rapidly growing discoveries of astronomy and astrophysics, of the exciting discovery of new planets, and other findings in the field. These new methods may build an educationally strong framework that could benefit visual learners and provide life-changing experiences.

© The Author(s) 2017
T.F. Slater and C.B. Tatge, *Research on Teaching Astronomy in the Planetarium*,
SpringerBriefs in Astronomy, DOI 10.1007/978-3-319-57202-4_6

Although research has shown mixed results on the viability of using the planetarium as an educational tool, the results were based on old tools and methods. The new digital tools of the 21st century provide a wonderful opportunity to develop and improve the field with 3D spatial simulations, for example. Technological advances now allow people to take education research in the planetarium as a learning environment to a much deeper level than ever possible before.

Index

Printed in the United States
By Bookmasters